Rockets, Reactors and Computers Define the 20th Century

For- Carol -
Thanks for your
interest -
& my visit -

Rockets, Reactors and Computers Define the 20th Century

A Firsthand Look at the Century's Most Pivotal Innovations

Charles Bradshaw

Charles L. Bradshaw

Providence House Publishers
WWW.PROVIDENCEHOUSE.COM
FRANKLIN, TENNESSEE

TENNESSEE HERITAGE LIBRARY

Printed in the United States of America

11 10 09 08 07 1 2 3 4 5

Library of Congress Control Number: 2007938543

ISBN: 978-1-57736-396-5

The image used to open each chapter depicts Janus, Roman god of beginnings.

Cover and page design by Joey McNair

Images on the dust jacket depict landmark innovations and personalities in the fields of rockets, reactors, and computers. The photos of the Saturn V leaving the launch pad and Dr. Wernher von Braun viewing readings at take-off are courtesy of the U.S. Space & Rocket Center in Huntsville, Alabama. The photos of Robert Oppenheimer and of the Shippingport Reactor, the first full-scale nuclear power generating station in the United States, are courtesy of the U.S. Department of Energy. The ENIAC computer image is a U.S. Army Photo; the photograph of computer pioneer John Mauchly is from the John W. Mauchly Papers, Rare Book & Manuscript Library, University of Pennsylvania, used by permission.

PROVIDENCE HOUSE PUBLISHERS
238 Seaboard Lane • Franklin, Tennessee 37067
www.providencehouse.com
800-321-5692

Dedicated to the memory of the beautiful Bernice,
mother of Carole, Bob, and Al

And in honor of
Helmut Hoelzer,
my mentor and good friend

Contents

Preface

The three defining components of the twentieth century arose out of the decade of the 1940s. The time for these events had come, and they were hurried along by World War II, which gave the space age, the nuclear age, and the computer age official births. The decade of the 1950s saw the push to understand and use nuclear technology for the good of mankind; the use of rockets to place satellites and man into space; and the use of the computer to accelerate research, revolutionize the way business is done, and increase the tempo of everyday activities.

Modern rocketry and space flight began at Peenemünde on the island of Usedom in the Baltic Sea on October 3, 1942. Here, Dr. Wernher von Braun and his rocket team—following the work and ideas of professors Robert Goddard, Hermann Oberth, and others in the 1920s and 1930s—designed and launched the first of the guided missiles they were to build during the war. Von Braun and many of his group moved to America after the war with hopes to continue working on their dreams of sending man into space and eventually to the moon. The cold war began, and America needed weapons again. The von Braun engineers were the only experienced rocket builders in America, and they were enlisted to lead America's guided missile program in the early 1950s and the way into space with *Explorer 1* in January 1958.

The nuclear bomb which exploded over Hiroshima, Japan, on August 6, 1945, officially opened the nuclear age. The bomb was built by the Manhattan Project led by Dr. Robert Oppenheimer and Gen. Leslie Groves. They were assisted in this herculean effort by more than one hundred top nuclear scientists from throughout the world. The 1950s were the years when many scientists expressed regrets at having launched the nuclear age while others studied ways to control the genie out of the bottle and how best to use this tremendous new source of energy.

The modern computer age began during World War II with the development of the Mark 1 computer at Harvard, the ENIAC at the University of Pennsylvania, the Zuse in Germany, and the Colossus in England. These computers were used to solve problems associated with building and using instruments of war. UNIVAC, a descendant of ENIAC, was introduced in 1951 as the first computer for sale, and stimulated the interest of many electronic firms in the anticipated demand for these new "giant brains."

The Great Depression came to the United States soon after our family had moved from Knoxville, Tennessee, to the zinc mining town of Mascot, located some twelve miles north of Knoxville on the Holston River. I was six years old and ready to begin school. Mascot was as good a place as any and better than some in which to grow up during the Depression since everyone worked for the same company, the American Zinc Company, and knew all their neighbors. Surviving hard times soon became a community project. The company owned all the houses, the commissary, and everything else in town, and did as well as it could to help the employees through the days of no work and no income. The local doctor, also on company payroll, repaired broken bones, handed out medicine for the flu, and brought babies safely into the world. The company divided up the work so that each employee had one or two shifts a week to avoid going completely broke, and when American Zinc ran out of cash, it paid the employees in scrip redeemable for groceries at the commissary.

The only social events in our town were at the Baptist church or the community center, and the entertainment was always music of some kind. In the church, the music usually consisted of hymns sung by all, but sometimes a quartet with peppy songs would get some feet to stomping. At the community center or the schoolhouse, and many times on someone's front porch, a local string band consisting of a banjo, a guitar, a fiddle, and maybe a French harp (harmonica), would play tunes like "Foggy Mountain Breakdown," "Ida Red," and "She'll Be Coming 'Round the Mountain" and there was always someone who would try to sing.

Much of the summer was spent preparing to survive the winter. Large gardens furnished beans, potatoes, tomatoes, peanuts, cabbages, cucumbers, corn, pumpkins, and okra to be preserved by drying or canning. The cabbage would be turned into sauerkraut or buried under a large dirt mound to be dug up during the winter. The surrounding fields and woods were searched for blackberries, walnuts, chestnuts, chinky-pin nuts, and scalybark hickory nuts to be gathered and saved for the winter.

Some families saved up a few dollars to purchase young hogs, called shoats, which were kept through the summer and fattened up for hog-killing time, which came soon after Thanksgiving. After everyone had eaten all the fresh pork they could stand, the rest was salt-cured or cooked and canned.

There was still some free time to run barefoot over the pastures and through the woods and to catch some catfish in the Holston River.

In late August, every family faced the challenge of getting a pair of shoes for each child and finding money to buy their school books. A good community spirit kept everyone fed, clothed, and most important, in school and in church.

In the mid-1930s, some families purchased radios on credit and all outside activities were scheduled around *Amos 'n' Andy*, *The Jack Benny Show*, the Joe Louis fights, the World Series, and the *Inner Sanctum*. President Roosevelt delivered his *Fireside Chats* via radio to homes, where all the neighbors gathered and listened to messages that would always be upbeat and promise that times would get better. The Philco, Zenith, and Atwater Kent radios provided a path to the outside world.

The Great Depression had one—and perhaps only one—beneficial effect on the people who experienced it. Those years prepared the young men and women of the nation to accept hardships and to cope with problems and obstacles they encountered in the war years of the 1940s and the great expansion years of the 1950s. A rough ride in later years was met with determination and optimism.

The United States entered World War II the year I graduated from high school, and soon I was on my way into the navy, where I would see more of the world than I had ever known existed. After the war, I was able to finish college on the GI Bill and to spend a few years teaching in college. I then spent three years working the nuclear problem at Oak Ridge, seventeen years with Dr. Wernher von Braun and his group of rocket scientists, and some fifty years working with computers. During the ten-year period of 1951–1961, I was fortunate to work under the supervision of some outstanding scientists and engineers. Dr. Hans Friedrich, a university professor in Germany before joining von Braun in the German V-2 missile project, introduced me to rocket science and gave me a good foundation in the fundamentals. Dr. Adolph Thiel, also a von Braun team member and an outstanding rocket engineer, taught me some of the engineering aspects of missile construction and flight. Dr. Rudolph Hoelker, an early member of the V-2 missile project and an expert in numerical procedures, gave me a good foundation in the numerical solution of ordinary differential equations during the three years I worked with him.

I worked in the mathematics panel at Oak Ridge for three years and had the good fortune to work under the supervision of Dr. Alston Householder. At Oak Ridge, the emphasis was on the numerical solution of systems of partial differential equations related to the design of nuclear reactors. In the early 1950s, Dr. Householder became the best known applied mathematician in the United States, and his advice was solicited from throughout the world. For many years, I was deputy to Dr. Helmut Hoelzer, a genius in electronics and the inventor of the electronic analog computer. Dr. Hoelzer, working with Dr. Ernst Steinhoff, was successful in developing electronic guidance technology for the early V-2s at Peenemünde.

During my seventeen years as a member of the von Braun rocket team, I was able to observe close up and learn from Dr. von Braun, Arthur Rudolph, Ernst Stuhlinger, Chuck Lundquist, and many other outstanding scientists and engineers.

From the perspective of these years, I have related some of the events leading to the technical, civil, and cultural revolution caused by the splitting of the atom, the propelling of rockets into space, and the advent of the digital computer. This book is not intended to be a comprehensive, precise, and extensively researched history in any of these three areas. Instead, it is hoped that the three major developments, which took place during the 1940s and 1950s and set the course for future years, are presented in a way which can be understood by the nonscientist with an interest in the change brought about by these defining innovations of the twentieth century.

My intention has been to provide enough detail so that the careful reader can get a general understanding of how the atomic bomb was made and exploded; how a nuclear power reactor is constructed and produces useful energy; how a guided missile moves through space; and how computers work and have become a controlling element in our society.

I have included a few symbols from algebra, trigonometry, calculus, and physics to perhaps motivate the students of today and to assure them there is an application for the abstract things they study. The reader who is frightened by equations and math symbols can skip over these without loss of continuity. However, for the more adventuresome, a careful study of these formulas is recommended since I have put some thought into presenting them in a manner so that they can be understood by those with no advanced training in mathematics or engineering. This does not include the equations and formulas on pages 153–156 or 203–205, which are given for illustration purposes on how it was done in the early days.

Along the way, I've included a few personal stories about my friends and colleagues to better show the personal side of this period of technical revolution and change, and perhaps preserve additional information about the people who helped shape the last century.

ACKNOWLEDGMENTS

In gathering together the papers and documents from the 1950s and reviewing the events of that era, I am reminded once again how fortunate I have been to be associated with the wonderful men and women who conceived and developed the guided missiles, the nuclear reactors, and the computers which define the twentieth century. I want to thank them for the really good experience, but most of all for their friendship.

I have again called on old friends for help, and as always, they were ready to give assistance. Chuck Hubbard has answered questions and written a good description of the manner of operation and engineering of the ORACLE computer we used in the early 1950s. Joe Brake, Grady Holt, Kitty Holt, Lee Stulce, and Stacey Mott have read all or parts of the book and have made valuable comments. I thank them very much and I have enjoyed renewing good friendships of the past.

Dave Palmer, Frank Kyle, and Denny Burnham, who were my colleagues at Vanderbilt University, read the manuscript and met with me on two occasions. They went through the nuclear physics, the mathematics, and computer science to check some of the technical details. I worked with these good men for nearly twenty years, and it was a pleasure to sit with them again and discuss technical matters.

Patricia Brake, an author and professor from Knoxville, and Lisa Huddleston, my brilliant neighbor, read the entire manuscript and made many corrections in punctuation and sentence structure. I learned that mathematicians can be too interested in symbols and shorthand and not enough in communication. Patricia and Lisa were able to get the manuscript in condition to present it to a publisher.

My wife, Loyce, who is my best friend, supporter, and critic, encouraged me to start this book and to see it through to the end. She has read the manuscript and made valuable suggestions for clarity of presentation.

Prologue

The twentieth century opened with transportation, living conditions, and the everyday way of life much as they had been for the previous three hundred years. The horse and buggy was still the principal mode of transportation and most people still lived a rural life where farming with hand tools and a mule was the primary occupation.

The great inventions of the printing press, cotton gin, and steam engine caused an awakening of what might be possible in the future, and already the steam engine showed promise of giving aid to the mule in the form of a tractor. The printing press offered more information to individuals who were beginning to take an interest in things outside their own towns, and soon, the Sears, Roebuck and Co. catalog would bring the department store to the home. The manufacture of cloth goods became more common, and cotton mills moved south for closer proximity to the cotton fields. No one could, however, have predicted the amazing changes to take place in the coming years.

The first forty years of the twentieth century witnessed many exciting developments, primarily in the form of transportation and power machines for both the farm and the home. The combustion engine was proven workable, and its application began with automobiles, farm tractors, and the airplane. The automobile became affordable to many when Henry Ford introduced the Model T and then followed with the Model A. Other companies which built cars with names such as Plymouth, Chevrolet, Dodge, Studebaker, Cadillac, and Packard were formed or expanded, and suppliers such as Western Auto and Goodyear came into being to sell motorists tires, parts, and accessories for their automobiles.

As the conversion to the combustion engine took place, the tractor became an essential and affordable tool for the farmer. In 1893, Fred Maytag formed a company in Iowa to build implements which would make the tractor a more effective and safer farm aid. The effort of Maytag led to a major consumer industry, which developed into areas such as the home washing machine and other labor-saving devices for the home and farm. As manufacturing facilities increased, many farmers migrated to Detroit and other cities where industry was coming alive.

The airplane attracted global notice during World War I with the heroics of the Red Baron and Eddie Rickenbacker. During the 1920s, stunt pilots and daring airmail runs were reminiscent of the pony express and the rough-and-tumble cowboys of the 1850s West. Airplanes piloted by these air cowboys started showing up at county fairs and would take passengers with a few dollars and a lot of nerve for a ride above the cow pastures. In 1927, Charles Lindbergh firmly established the airplane as a transport vehicle with his solo flight across the Atlantic Ocean from New Jersey to Paris. In the 1930s, Howard Hughes and others envisioned the airplane as a mode of international travel as Trans World Airlines and Pan American Airways began to circle the earth.

The Great Depression, coming to the United States in 1930, prevented greater industrial expansion. It's amazing that companies such as Ford, Chevrolet, Plymouth, and Maytag were able to do as well as they did when the citizens had so little money to spend. The innovation of paying for major purchases over time (time, or on credit, payments) came along at just the right time and the introduction of electric power to rural areas had a major impact on the use of home appliances.

The mid-twentieth century was ready to explode with innovation and change; technology was poised to impact the lives of people everywhere, and the world was on the verge of seeing the practical deployment of the three greatest innovations of this—and perhaps any—age.

Chapter 1

From Tennessee Central to the Big Spring

In March 1951, Cookeville and the rest of the Cumberland Plateau in the eastern portion of Middle Tennessee were recovering from the worst winter in the twentieth century. The blizzard of 1951, which started on January 29 with a heavy, freezing rain leaving a sheet of ice on all trees, streets, and byroads, took down most of the electric power lines. The ice storm was followed by ten to twelve inches of snow and temperatures which dropped to between twelve and fifteen degrees below zero. The county was able to salt the main streets but the freeze was so hard that the salt was no help, and there was no snow removal equipment to clear the roads.

Cookeville was a college town and had been populated with veterans of the Great Depression and the war; some who had marched across Europe with General Patton and others who had fought their way through the jungles of the South Pacific. These people had learned how to handle a crisis and survive so there was no panic about the Cookeville ice storm; the strong and the capable made sure those who were not were given assistance.

Bernice and I were in our fifth year in Cookeville, having come there at the conclusion of a honeymoon which had been shortened by my desire to get back to college. I had served three years in the navy during World War II and was, like millions of other returning veterans, anxious to get on with my career, whatever it might turn out to be. I majored in mathematics because it was my easiest subject and the shortest route to a college degree. I finished a bachelor of science degree in mathematics in one year (having had some college before the navy and some during) and taught part time while finishing a master of arts degree at the University of Tennessee.

Our first child, a beautiful little girl whom we named Carole Sue, had been born on July 4, 1947, and now Bernice was seven months pregnant with our second child, a son whom we were to name Charles Robert.

I was finishing my third year as an instructor at Tennessee Tech and would be promoted to assistant professor with tenure at the end of the academic year. Tennessee Tech, then known officially as Tennessee Polytechnic Institute, was undergoing major innovations and growth, and was enjoying a booming postwar expansion in facilities and student body under the leadership of President Everett Derryberry.

One day in mid-March after completing my morning classes, I was driving home for lunch. We felt fortunate to have a car (1937 Pontiac) since production of automobiles had stopped during the war, and all the returning veterans placed a heavy demand on available cars. The car had a Truetone radio and, as usual, I was listening to country music on WHUB, the radio station which called itself the Hub of the Upper Cumberland.

Stacey Mott, my good friend and fellow math major, was the announcer, and he played a song by Hank Williams. It was a sad song, as were most of Williams's songs, about a wooden Indian named Kawliga who stood outside the barber shop in the rain and the snow and, worst of all, loved a maid but lost her.

At the conclusion of the song, Stacey took time out to read a public service announcement from the Department of the Army which was opening a new research center in Huntsville, Alabama, projected to be critical for our future national security. Personnel recruiters were coming to Cookeville the very next Saturday to recruit scientists and engineers to become part of this important project. The recruiters were to set up an office in the Shanks Hotel in downtown Cookeville.

The announcement, however, went right over my head as I continued to think about the improvement in the weather and the good bass fishing that was only weeks away at Center Hill and Dale Hollow Lakes.

On Saturdays, we always took a trip into downtown Cookeville to do the little shopping we could afford and to buy Carole a trinket. We purchased the week's groceries from Mr. Foutch, the owner of a grocery and meat market who would extend credit to us until our monthly paycheck of about $280 came through.

At the end of the morning, we would go to the depot to see the noon train, which made a stop in Cookeville. Meeting the noon train on Saturdays and Sundays had been a tradition in Cookeville for almost fifty years.

The Tennessee Central Railroad (TC) offered passenger service from Harriman to Nashville, and ran right through the western part of Cookeville where the depot became the center of activity and business development.

The Tennessee Central began serving east-west passengers and transporting freight in 1890. Before that, there was no rail service, or any other service for that matter, through the highlands. The TC, as it was affectionately known, came into being due to the exhaustive efforts of a Nashville businessman named Jere Baxter and Alexander Crawford, a Pennsylvania iron manufacturer who was drawn to the area because the Cumberland highlands were filled with large seams of bituminous coal.

The first link of the railroad connected Cookeville to the already existing tracks at Lebanon, some fifty miles to the west, and then east some twenty miles into Monterey, where the richest seams of coal had been found. Over the next few years, the track was extended from Monterey to Crossville, then down the steep mountain to Rockwood, and then into Harriman where it connected with north-south trains of the Southern and L&N Railroads and eastward into Knoxville on the

Southern. From Nashville, the rails eventually continued into Hopkinsville, Kentucky.

The TC provided a way for farmers on the plateau to get their products to market in Nashville and for merchants to receive goods from the east or west; they had ordered their goods from drummers, who now rode the TC to Cookeville and set up shop in the Shanks Hotel. One day each week, the TC would come equipped with a freight car to carry chickens to Nashville, and on another day, it would bring ice so it could transport eggs. The passenger service provided a way for the ladies of Cookeville to go to Nashville for a day of shopping at Harvey's, Castner Knott, and Cain-Sloan. The trip from Cookeville to Nashville took more than three hours, so the shopping trips, which became a major social event for the Cookeville ladies, took a long and tedious twelve hours.

On some Saturdays, we would see country music stars such as Roy Acuff, Homer and Jethro, and Archie Campbell from Bulls Gap on the train as they headed from their weekday jobs in Knoxville to Nashville for a performance of their Saturday night jobs at the Grand Ole Opry. Often a country music performer would come out to the platform at the end of the train and play a song for the crowd.

I had grown up listening to the kind of songs they sang and had often stood on the sidewalk of Gay Street in Knoxville to watch the performances on the *Midday Merry-Go-Round.* This popular live show on radio station WNOX drew a crowd every day during the lunch hour where spectators could stand on the sidewalk and see the show through the large plate glass windows or pay ten cents to go inside and get a seat in the small auditorium. Lowell Blanchard hosted this popular radio broadcast, and many country music entertainers such as Chet Atkins, Archie Campbell, Homer and Jethro, and Roy Acuff got their start in country music on this program.

My grandmother, whose maiden name was Mattie Atkins, had made the trip to Knoxville on the train, and paid a dime to go inside and watch her cousin Chet perform on the *Midday Merry-Go-Round*. It was very unusual for my grandmother to leave her home in Powder Springs, even for one day, and it was more unusual to see her pay a dime for something other than food or cloth.

As we were leaving the train station, I happened to notice the Shanks Hotel just across the street. It was a three-story building spread across half the block facing the train depot from the south. I remembered the public service announcement Stacey Mott had read on WHUB a few days earlier, and thought to myself, *Why not?* I briefed Bernice and suggested that she and Carole go on to the ten-cent store, where I would meet them a little later. Carole was very happy since this would give her more time to look in the store.

The desk clerk at the Shanks Hotel informed me that the personnel recruiters from Huntsville were on the second floor and they had not had very many job applicants. The recruiters seemed glad to see me, and we got right down to the business at hand. As we discussed the courses I had taught and taken, they referred to a list, and I could tell when they got a match. The recruiters were particularly excited when I mentioned teaching courses in applied differential equations and the theory of equations. They made several notes and then asked about my salary requirements.

I told them I had just been promoted to assistant professor in mathematics and I expected to be raised to thirty-three hundred dollars a year. With my tenure assured and having the pleasure of living in a small college town with good bass fishing nearby, it would not be advisable for me to move for less than five thousand dollars a year. This figure seemed like easy street to me, and Bernice and I could do all we ever hoped to do with a raise of 50 percent.

As I left, the recruiters assured me they would fully evaluate my application and be in further correspondence with me. As I rushed off to find Bernice and Carole at the ten-cent store, however, I doubted there would ever be a job offer since my salary request was so out of range.

A few weeks later, I was again on my way home for lunch, this time thinking about Honky Crouch, my good fishing buddy. He recently had gone home to Viola, a small town near McMinnville in Middle Tennessee, and bought a used outboard motor. We had gone fishing twice in April and were planning another trip in early May, which would be our last for a while since Charles Robert's delivery date was during the third week of May. After classes that afternoon, Honky and I planned to mount the motor in a barrel full of water to see if we could get it started. There were no reliable or easy-to-start outboard motors built in this postwar era. Many companies manufactured them in what appeared to be a race to see who could build the first good one. Outboard motors carried the names Martin, Firestone, Goodyear, Wizard, Elgin, and the oldest of all, Evinrude, which was considered to be the best. The motor Honky bought was a Martin, purchased for thirty-five dollars, and I wondered if we would ever get it started.

As I pulled into our gravel driveway, Bernice and Carole were waiting for me on the front porch of our rental house with a surprise. I had received a special delivery package from the United States Army in Huntsville. The enclosed letter stated that if I would complete the forms enclosed, and if things checked out, they would offer me a job as a "Mathematician GS-10" at five thousand dollars a year! In a few weeks, after our son was born and the academic quarter was over, we placed all our belongings in a borrowed trailer, and with our new son, Bob, and Carole, and her entourage of dolls and imaginary playmates, drove our 1937 Pontiac through Tennessee towns with names including Sparta, McMinnville, Tullahoma, Mulberry, Lynchburg, and Fayetteville. We drove slowly through Lynchburg and marveled at the large buildings on hilltops full of Jack Daniel's whiskey aging away to be sipped at some future time.

In Huntsville, we spent Saturday night in a motel on the edge of town. Early on Sunday morning, we went into downtown Huntsville for our first visit and Bernice asked that we locate the Presbyterian churches. There were two, and Bernice picked the

Central Presbyterian Church as the one we would attend since it subscribed to the same brand of Presbyterianism we had practiced in Cookeville.

The city square was typical of a town the size of Huntsville. The courthouse stood in the middle, surrounded on all sides by the main businesses in town. On the west side of the square was a group of small buildings and offices all huddled together and bordered by a large structure housing the First National Bank. A man sitting on a bench nearby told us that side of the square was called "Cotton Row," and it was where men got rich or poor trading cotton futures. From the looks of the offices, we assumed most got poor. We soon learned that Madison County grew more cotton than any other in Alabama, and that very successful cotton brokers occupied those little offices huddled along the bluff.

The First National Bank stood majestically on the southwest corner and with its large marble columns gave every indication of strength, integrity, and durability. The bank was built on a rock bluff, from under which a giant spring of water flowed, forming a big pool. This was the largest spring we had ever seen, and we later learned that Huntsville was sometimes known as the City of the Big Spring. From this big spring, crystal-pure water flowed out across a large field covered with watercress. The watercress was harvested by the owner of the field and shipped internationally, leading some to call Huntsville the "Watercress Capital of the World."

One of the very first white settlers in all of North Alabama was John Hunt, who made his home at "the Big Spring." The exact date of Hunt's arrival is unknown, but it was sometime during the year 1805. The land around the Big Spring was the shared hunting ground of the Cherokee and the Chickasaw, and there was a good supply of deer, bear, geese, and rabbit. The story is told that Hunt spent most of his time killing rattlesnakes which inhabited the bluff behind the Big Spring, so he had little time left for hunting the wild game nearby. Due to the intensity of his rattlesnake hunts or perhaps by simple oversight, Hunt failed to register his squatter's claim, and he lost the land he had hoped to claim for his own.

Other pioneers soon arrived, as treaties with the Indians opened up more land for settlement. Madison County was established in 1808 by the governor of the Mississippi Territory. A Georgia planter named Leroy Pope saw the potential of the valley surrounding the Big Spring and purchased the land for twenty-three dollars an acre; he named the town Twickenham.

In 1811, the town was incorporated and renamed Huntsville (the early settlers never liked the English-sounding name Twickenham). This renaming may also have occurred to make up for the injustice many of the community leaders felt had been done to John Hunt. An illiterate pioneer woodsman, Hunt had hoped that a benevolent government would express its gratitude to him by an outright grant of the Big Spring land he had settled. When this did not happen, he disappeared in 1811, never to be heard from again.

Huntsville soon became one of the most promising little communities on the frontier. In 1819, a constitutional convention met in Huntsville to elect William Wyatt Bibb as the first governor of the newly constituted state of Alabama. Madison County became a major cotton producer, using the nearby Tennessee River to ship cotton to New Orleans and other cities along the Tennessee, Ohio, and Mississippi Rivers.

To learn more about our new hometown, we moved to the south side of the square where a large mercantile store stretched almost the entire length of the block as a dominating business. The building was two stories with a glass front, and as we walked along the sidewalk, we could see clothing, furniture, food products, hardware, appliances, and a lot of other items a growing family would need. We knew we would be shopping here so I backed up in order to see the name of the business. In large letters near the top of the building, the sign read *"T. T. TERRY—ON THE SQUARE."* I read the message and interpreted the double meaning of the phrase "on the square." That was good advertising.

There was another sign just below the name of the store and in letters almost as large. I had to walk out into the middle of the street to read the whole sign because it covered the length of the square. This sign read *"GREAT IS THE POWER OF CASH."* I read and reread this phrase, looking for a double meaning, but I couldn't get the double meaning if one existed. This was *not* good advertising.

Discouraged, I muttered just loud enough for Bernice to hear, "We're gonna have a helluva time getting any credit down here."

Bernice gave me a serious look and said, as she did so many times in our lives together, "Don't worry, everything will be all right." She added, even more seriously, "Don't use that word in front of the children."

We walked to the east side of the square and found two places we would be going to on a regular schedule. One was Wimpy's Hamburger Palace, where hamburgers cost ten cents, and the other was the Crescent Drug Store with an old-fashioned ice cream counter that served large ice cream cones for five cents.

On the north lawn of the courthouse, placed in a conspicuous spot, was the statue of the ubiquitous Confederate soldier which fronted most courthouses in the Deep South. The soldier had his rifle at the ready position, pointed toward the North, just in case they came back again. As we passed, I looked into the eyes of the Confederate soldier and told him, "Get ready. The Yankees are on their way again."

Chapter 2

How I Became a Rocket Scientist

Redstone Arsenal is an expansive reservation consisting of thirty-eight thousand acres stretching along the Tennessee River just south of Huntsville. During World War II, there were two separate installations on the reservation, one known as the Redstone Arsenal and the other as the Huntsville Arsenal. These installations were major manufacturing sites for ammunition and related explosive weapons. Each consisted of many buildings given wide berths due to the nature of the products being manufactured.

The personnel office for the new organization, known as the Guided Missile Development Division (GMDD) of the army, had established its offices at the headquarters of the old Redstone Arsenal some twelve miles south of Huntsville near the Tennessee River. I had been directed to report to these offices on a Monday in early June 1951 to be interviewed for a job assignment.

After filling out a few forms, I was escorted to a van and taken to the area where the operating elements of the GMDD were located on the northeast side of the arsenal. The area was fenced, had a guard gate, and consisted of three two-story office buildings and two or three smaller buildings. The driver of the van told me that there were about 120 German-born scientists and one hundred civil service employees and military personnel who were then employed in the GMDD, and that the little complex where I was to work was named "Squirrel Hill."

I reported to an office on the second floor of Building 111. The door to the office was closed, and when I knocked gently on the door, I heard a grunt from inside, which I interpreted as an invitation to enter.

On entering, I saw a very serious man wielding a slide rule and looking at a large sheet of paper on his desk. I immediately assumed I was about to meet my first rocket scientist, who motioned for me to wait as he completed the series of calculations he was doing. I stood more or less at attention with the folder I had brought tucked under my arm, and it was awe-inspiring for me to stand and watch the first rocket scientist I had ever seen hard at work.

When the rocket scientist completed his series of calculations, he wrote down a couple of numbers and turned his attention to me. He asked, "*Was haben sie das unter dem arm?*"

Wow! I thought, *These rocket scientists sure have funny accents.*

I handed him my folder, and he proceeded to look it over and gave it a couple of grunts. After a minute or so, he looked at me and asked, "Mr. Bradshaw, do you truncate?"

Now, I wasn't quite sure whether that was a technical question or a question about my morals, so I decided to gloss over my answer with a grunt and a mumble.

The rocket scientist looked at my paper again and then asked another question, "Mr. Bradshaw, do you root polynomials?"

I was ready for that question as my mother had once given me some advice which would help. In the mid-1930s while we were living in the zinc mining town along the Holston River in East Tennessee, I observed the dead-end cycle taking place about me. When a boy reached sixteen, he would marry a girl from a street or two away, and the couple would move in with one of their parents. The boy would apply for a job with the mining company, and in a year or two, he would have a job in the mines and a company house of his own. The children became copies of their parents and grandparents.

Having read books and stories by Rudyard Kipling, Jack London, and Mark Twain, I had made some big plans for my life, and the cycle I saw then did not fit those plans.

One day I told my mother what I had observed and that I was not going to follow the town's cycle. My plans were to go out in the big world, out beyond Knoxville, and see what was there.

Mother sat me down and talked rather sternly to me. She said, "Son, you are a good boy, and you need to stay with us." She continued: "We have had some pretty hard times here in our town but we have always had enough to eat, and you are growing up strong and healthy. Out there in the big world where you are talking about going, it is *dog eat dog and root hog or die*."

Remembering the words of my mother, I assured the rocket scientist I could root polynomials or anything else that might need rooting.

This seemed to satisfy the rocket scientist, and he led me down the hall to an empty office and sat me down in front of my first computer. It wasn't really a computer but a Friden electromechanical adding machine which could, in addition to its basic functions of adding, subtracting, and multiplying, divide if you had time to wait.

I sat down and began the process of learning to operate the Friden calculator and hoped I was on my way to becoming a rocket scientist. My boss gave me some arithmetic calculations to do on the Friden so the results could be compared with the results he obtained on the slide rule. It took me longer, but my calculations were correct to six decimal places, while only one or two decimal places were good on the slide rule.

In a couple of days, my boss told me I was moving to another office at the end of the hall, and I would be working for another rocket scientist. My new boss was Dr. Hans Friedrich, who had been a university professor in Germany before joining the von Braun group in the rocket development program at Peenemünde.

I had been assigned to him as a student to learn aerodynamics and flight mechanics. It soon became obvious that Dr. Friedrich was an outstanding teacher. I was to be taught in the old-world style of sitting at the professor's feet in one-on-one learning sessions. My desk was placed directly in front of Dr. Friedrich's so we faced each other all day.

Each morning I would move my chair beside him, and he would lecture while making notes on a pad. After about an hour or so of this, I would take the notes back to my desk and spend the rest of the morning studying and transcribing them into the form of a technical paper. In the early afternoon, I would move my chair back beside Dr. Friedrich, who would go over my paper to determine if I had captured and fully understood the points of his morning lecture. We would then go on to the next subject, and the process would be repeated.

We continued in this manner for two or three weeks. When I had a question all I had to do was to raise my head and ask. This turned out to be the most productive learning situation I was ever to experience.

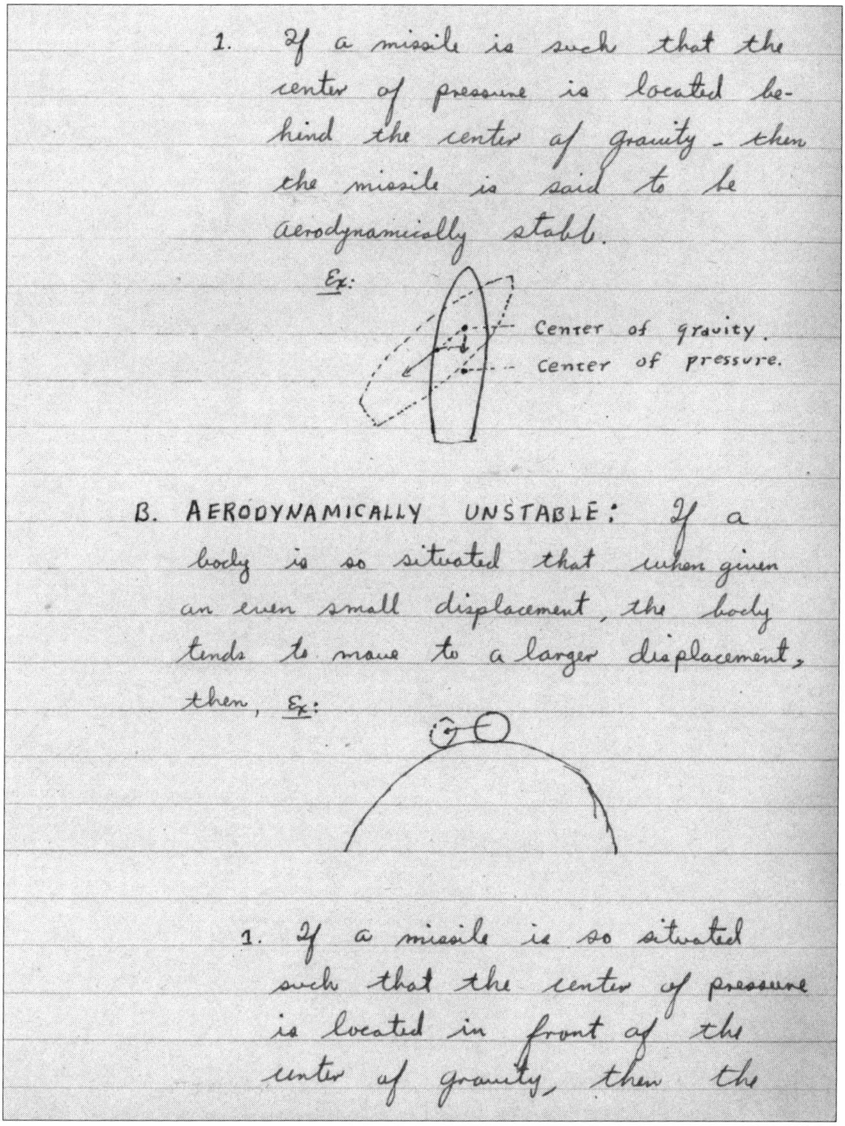
1. If a missile is such that the center of pressure is located behind the center of gravity – then the missile is said to be aerodynamically stable.

Ex:

B. AERODYNAMICALLY UNSTABLE: If a body is so situated that when given an even small displacement, the body tends to move to a larger displacement, then, Ex:

1. If a missile is so situated such that the center of pressure is located in front of the center of gravity, then the

Above and on page 20: Notes from the lectures of Dr. Friedrich, 1951.

June 14, 1951

C. ATMOSPHERIC FORCE: This force is divided into two components – in the direction of the flight and normal to the flight – known as drag and lift.

C_L = lift coefficient.
C_D = drag coefficient.

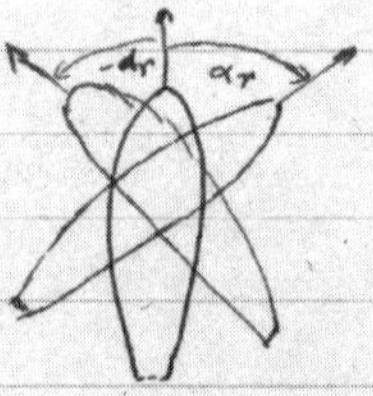

$$C_D = C_{D0}^{*} + C_{Di}\,\alpha_r^2$$ ~~(defn.)~~

logic reason & ch. by exp.

$$C_L = \frac{\partial C_L}{\partial \alpha} \cdot \alpha_r$$

$$= C_L' \cdot \alpha_r$$

The drag is that force acting along the ~~axis~~ flight path of the missile.
The lift is the aerodynamic force acting in a direction normal to the flight path.

$^*C_{D0}$ – the drag coeff. for a missile whose air flow is parallel to the longitudinal axis.

Dr. Friedrich started with the most fundamental concepts of rockets. Jet propulsion is a means of locomotion whereby a reaction is imparted to a device by the momentum of ejected matter. This is called duct propulsion which means that surrounding fluid is ducted through the device and accelerated to a greater momentum, by mechanical or thermal means, prior to ejection. There are three types of duct propulsion:

1. Turbojets: The working fluid is compressed by mechanical means, burned, and expanded in a nozzle. This is the most common type of rocket.
2. Ramjet: The medium is air which is compressed through the speed of the vehicle.
3. Pulse Jet: Intermittent; the medium (air or fluid) is repetitively admitted, compressed, heated, and expanded.

The concept of stability is most important in rocket design. Dr. Friedrich explained that an aerodynamically stable rocket, even under a significant displacement, will tend to return to the stable position. In an unstable rocket, a minor move will cause the rocket to become more unstable and to perhaps destroy itself.

In determining stability, there are two important points in the rocket which must be calculated: the center of gravity and the center of pressure. Dr. Friedrich discussed procedures involving some knowledge of physics and calculus which could be used to calculate the center of gravity and the center of pressure. The center of gravity is the geometric center of the weight of the rocket and is where all the weight of the rocket can be considered to be concentrated. Likewise, the center of pressure is where all the pressure on the rocket is concentrated.

Measuring from the nose to the tail, if the center of pressure is located behind the center of gravity, then the rocket is stable; otherwise the rocket is unstable.

The forces acting on the rocket in flight were the next subject Dr. Friedrich discussed. The three phases of flight for a rocket are the ascension or the propelled phase, free-flight, and the die or dive phase. In each of these three phases of flight, there are different principal forces acting on the rocket.

The main force acting on the rocket during the ascension phase is the thrust or propulsion, which is the upward force on the rocket caused by the expulsion of gases from the engine. Thrust is measured in pounds or kilograms and has two components: propulsion and momentum. It was emphasized that thrust is not a push; it is a reaction. A rocket would operate best in space since the resistance to the reaction would be negligible, and the rocket needs nothing to push against. Good examples of reaction are the kick of a shotgun or a fire hose which, when turned on suddenly, can sit a grown man on his bottom. A differential equation derived from a fundamental law of Isaac Newton can define the thrust and its effect on the rocket.

A major force acting on the rocket while it is in flight is the force of gravity, which is dependent on three things: the force of gravity at sea levels (g), the average radius of the earth (R), and the instantaneous altitude of the rocket (y). Knowing these three parameters, it is then possible to calculate the instantaneous effect of gravity, at altitude y, on the rocket by the simple algebraic formula

$$g(y) = g\left(\frac{R}{R+y}\right)^2$$

Another force acting on the rocket in flight is the atmospheric force, sometimes called air drag or a resistance due to movement through the atmosphere. This force is divided into two components: one force in the direction of flight (drag) and one force normal (normal, meaning perpendicular) to the flight (lift).

The calculation of these factors requires a few steps in trigonometry and vector analysis. The atmospheric force would be 100 percent drag if the rocket is flying such that the direction of flight is exactly the same as the direction of thrust.

A factor which must be known to calculate the drag and lift components is the angle of attack of the rocket. The angle of attack is the angle between the longitudinal thrust axis of the rocket and the tangent to the flight trajectory. Again, some simple trigonometry can resolve the forces into the desired components.

Another number which is important is the Mach number (M) which is named for the Austrian physicist Ernst Mach. M is the ratio of the velocity of the missile to the velocity of sound. The speed of sound is about 740 miles per hour. If the rocket is traveling at a speed of $M<1$, it is said to be subsonic; if $M>1$, it is said to be supersonic; and if $M=1$, the rocket is said to be transonic. (Note: the symbol > means greater than and the symbol < means less than).

The flight path when a rocket is transonic is a critical period. Forces are acting during the transonic period which may cause difficulty in the flight and possible damage to or destruction of the rocket. In some instances during the transonic period, the center of pressure will move in front of the center of gravity causing instability and perhaps rocket destruction.

In order for the rocket to become a guided missile, it must have on board a guidance system with some basic intelligence.

The missile must have a means of changing course when given the proper command and of correcting disturbances which cause the missile to pitch, yaw, or roll. The early missiles executed the control required by using rudders, called jet vanes, which were placed in the jet stream and large movable fins on the sides of the missile.

The jet vane was placed exactly along the longitudinal axis and perpendicular to the planned flight path. The vane would be required to guide the missile from its vertical launch direction to the planned flight path, generally referred to as the alpha program. The jet vane changed course of the missile by deflecting the jet stream, very similar to the way in which the rudder on a ship deflects the thrust of the propeller and changes the ship's course.

The exterior fins could be moved by the control system to correct for disturbances which cause pitch, yaw, or roll. In later versions of guided missiles, the engines would swivel to execute the alpha program and to make flight corrections due to anomalies.

During flight, there are several forces acting on the missile which have to be considered as the range of the missile becomes greater and more accuracy is required. Some of these forces are wind, curvature of the earth, and the Coriolis effect (tendency for a missile in flight to drift sideways due to the rotation of the earth).

When the forces acting on the missile during the three phases of flight have been fully understood, the process of developing the equations defining the flight is relatively straightforward but requires a basic knowledge of calculus and vector analysis. These equations of motion are developed by applying another law of Newton, which generally states that for a body in motion, the force (F) exerted by the body is equal to the mass (m) of the body multiplied by the acceleration (*a*) of the body. In formula notation, the law is stated as: $F = m \times a$.

The most tedious task is analyzing all the forces acting on the rocket and resolving these forces into resultant forces along the three axes, *x*, *y*, and *z*. A good knowledge of trigonometry and vector analysis will get one through this step without too much difficulty. A fundamental knowledge of ordinary differential equations is required to write out the six first-order differential equations which define the path of the missile down the range to impact.

After I had been with Dr. Friedrich for about three weeks, he suggested I spend a few days with Dr. Adolph Thiel to study the engineering aspects of rockets. The time with Dr. Thiel was less structured than the time with Dr. Friedrich. We first discussed the three essential engineering components of constructing a guided missile, which were, in their order of engineering complexity and importance: the rocket motor, the missile guidance system, and the fuel tanks with associated plumbing and controls. Dr. Thiel gave me a few internal documents about the early days of rocket development and asked that I spend some time going through them.

As part of my tutorial with Dr. Thiel, I spent four or five days going through the papers provided on the history of rockets, made an outline of the information, and then went through my notes with Dr. Thiel.

The army was in the process of compiling a technical library containing many of the papers brought to the United States from Germany along with a few other books. These proved to be adequate for a cursory study of the history of rockets.

Rockets had been around for probably one thousand years, dating back to early use by the Chinese who deployed them in battles to start fires and harass their enemies. The Canadian Indians added rockets to their arsenal and used them against the British. Learning from the Indians, the British developed a military rocket and used it against the United States in 1814 in the unsuccessful attack on Fort McHenry outside Baltimore. (The United States' national anthem, "The Star Spangled Banner," contains the words "the rockets' red glare," as inspired by those rockets used in the attack.)

Rockets were commercially produced and sold primarily to seafaring vessels for the purpose of carrying a lifeline to shore in case of a shipwreck or for sending up a distress signal. These rockets were made of solid propellants similar to the sky rockets used during the Fourth of July and other celebrations.

In the nineteenth century, there were a few scientists who believed the rocket would be the vehicle to someday get man to the moon. A Russian teacher named Konstantin Tsiolkovsky (1857–1935) enthusiastically promoted space flight and wrote books on the subject as did the French author Jules Verne, who published *From the Earth to the Moon* in 1865. These men were a few years ahead of their time.

In the United States, the father of modern rocketry is acclaimed to be Dr. Robert Goddard, a native of Worcester, Massachusetts. An assistant professor at Clark University, Goddard began to experiment with rockets in 1915. His first experiment was to compute the efficiency of rockets by measuring the total momentum imparted to a pendulum by a fired solid-propellant rocket. Goddard reasoned that a rocket was a heat engine which converted heat into mechanical energy. His experiments led to a disappointing conclusion that only 2 percent of the energy created by the rocket motor contributed to the movement of the pendulum—representing a real problem with rockets.

Fortunately, this problem had already been met and solved by a Swedish engineer named Gustaf de Laval, who had developed a nozzle to meet the need for the most critical component in a machine which converts heat energy into kinetic energy. In his process of developing a more efficient steam engine, de Laval had developed a turbine whose wheel was turned by jets of steam. He found that the most efficient conversion took place when the nozzle was first narrowed (squeezed down), increasing the speed of the jet of steam to the speed of sound, and then expanded again. The nozzle, which today carries de Laval's name, creates the most efficient conversion from heat energy to kinetic energy.

Goddard experimented with the de Laval nozzle and produced jet velocities between seven thousand and eight thousand feet per second and efficiencies up to 63 percent. For rockets, the de Laval nozzle was the answer to the energy conversion problem.

Goddard did many more experiments with his ground-based rocket and proved, contrary to popular thinking at the time, that a rocket worked just as well in a vacuum as on the surface of the earth. Goddard also noted that the rocket was the most efficient heat engine with its 63 percent efficiency as compared with the steam engine's 21 percent and the diesel's 40 percent.

There was not a great interest in Goddard's work in the United States, and he was able to acquire only a limited amount of funding from the Smithsonian with which to build and launch some high-altitude rockets in an area near White Sands, New Mexico. He moved from the solid-propellant rocket and experimented with the liquid-fuel rocket which he had conjectured earlier simultaneously with Prof. Hermann

Oberth in Germany and Dr. Konstantin Tsiolkovsky in Russia. During the early part of the twentieth century, Goddard made several contributions to rocket development and subsequent space flight and is credited as being the first to accomplish the following:

- explored mathematically the practicality of using rocket propulsion to reach high altitudes and to go to the moon (1912)
- proved, by actual static test, that a rocket will work in a vacuum; that it needs no air to push against (early 1920s)
- developed and shot a liquid fuel rocket (1926)
- shot a scientific payload (barometer and camera) in a rocket flight (1929)
- developed a gyro control apparatus for rocket flight (1932)
- used vanes in the rocket motor blast for guidance (1932)
- received United States patent for idea of multi-stage rocket (1914)
- developed pumps suitable for rocket fuels (1922)
- successfully launched a rocket with a motor pivoted on gimbals under the influence of a gyro mechanism (1937)

Goddard, Tsiolkovsky, and Oberth were the first to envision the exploration of space using rockets. At the early part of the twentieth century, neither the culture nor the time was right in the United States or Russia to explore rockets or space. However, the politics, technical skills, and economy came together in Germany to form a perfect storm for the development of the guided missile.

In Germany, private companies had become interested in rockets and were funding experiments. Automobile companies such as Leyland and Opel attempted to set speed records using rocket motors, and when they used black powder for fuel, they created a major scene racing down a track. The novel by Jules Verne, *From the Earth to the Moon*, had captured the imagination of the youth in Germany and a movie about the woman in the moon (*Die Frau Im Mond*), based on the story by Verne and a book by Oberth, *The Road to Space Travel*, was being planned and promoted.

German universities were turning out brilliant and eager scientists and engineers who were anxious to use their skills in new adventures but these young people could not build or fly conventional aircraft due to the restrictions placed on Germany by the Treaty of Versailles at the end of World War I. Groups of amateur rocket enthusiasts formed, meetings were held, and the building of rockets was explored.

The best known of these groups was formed in Breslau in 1927 and called itself *Verein fuer Raumschiffahrt* (Society for Space Travel), using the initials VfR. The stated purpose of the VfR was to popularize the idea of space flight and to experiment with rocket propulsion. Oberth was named the president of the group, and Willy Ley, later to become a famous author of science and space travel in the United States, was a prominent member.

Two years after formation, the group had 870 members, including a seventeen-year-old recent high school graduate named Wernher von Braun. Wernher was the son of a well-to-do citizen who was the Minister of Education and Agriculture of the Republic of Germany.

Over the next few years, the VfR experimented with rockets, and having looked in all directions for financial support, was eventually able to acquire a building for constructing its rockets and a proving ground for testing.

In 1935, representatives of the German army visited with the group and explored the eventual use of the rocket as a weapon. This relationship eventually led to the German missile program, with the center of development and testing at Peenemünde on the island of Usedom, approximately 175 miles north of Berlin. The German army's development of the rocket as a weapon was placed under the leadership of then Col. Walter Dornberger. He selected Wernher von Braun, who by then had received a PhD at the University of Berlin, as the technical director.

During the next session with Dr. Thiel, the development of the rocket motor was the subject of interest.

The fundamental problems encountered in designing a rocket motor are proper injection of the fuel and oxidizer in the proper amount and mix, the timely and complete ignition of the mixture, the proper design of the combustion chamber and the exhaust throat, and preventing a burn-through in the combustion chamber or exhaust throat due to the extreme heat.

The development of an effective method of atomization and mixing of the fuel is a first priority. The von Braun group made significant improvements in the injection problem by using a mushroom head at the injection end with eighteen small centrifugal nozzles to more effectively spray and atomize the fuel and oxidizer in the combustion chamber. The result of shortening the combustion chamber from previous nozzle designs increased the exhaust velocity and the thrust of the rocket motor. A film of the alcohol fuel along the walls of the combustion chamber and the throat was proposed for cooling to prevent burn-through. A great deal of engineering over many months finally proved successful.

The next major engineering problem Thiel discussed was the development of a guidance system which would inject the missile warhead into a position and in an attitude from which to hit a target some 200–250 miles away from the launch site.

The ballistic missile was originally conceived by the German army as a replacement for long-range cannon. The guided missile was being thought of as a cannon with a barrel of some fifteen to twenty miles and an altitude of fifteen to eighteen miles. The guidance problem for the missile was to get the warhead to the injection point on the right course and with sufficient velocity to reach the target.

The guidance system of the early German missiles consisted of two main components. One was a sophisticated stabilized gyro system which would correct for variations from the flight plan in pitch, roll, or yaw. The gyro system would recognize an unplanned movement and send a signal to the jet vane or the outside fins for correction. However, a side wind which could move the whole missile and get it completely off course would not be detected by the gyro system. It was recognized that a guide beam would be needed to overcome this problem and to get the missile to the projected spot in space. It was soon determined that a series of guide beams would be needed to properly position the missile while in flight and this would require a mixing device capable of doing arithmetic, integration, and differentiation operations.

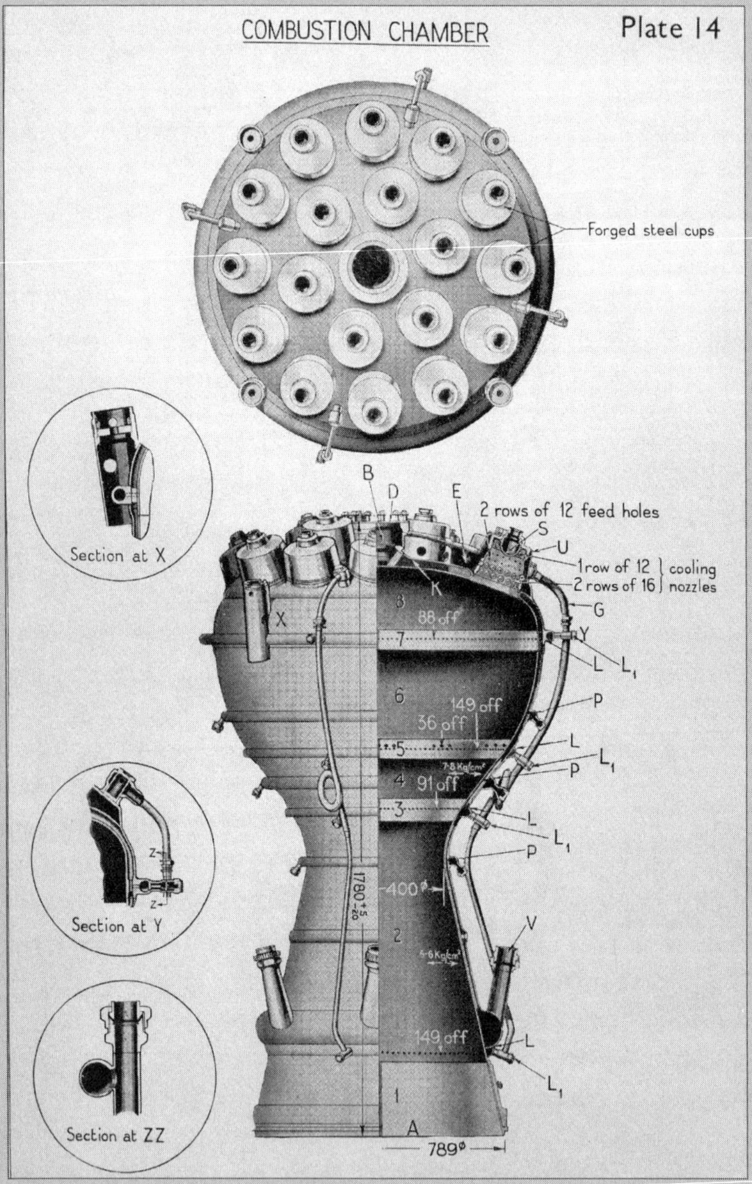

The V-2 rocket engine. Courtesy of National Air and Space Museum, Smithsonian Institution, (SI 79-13177).

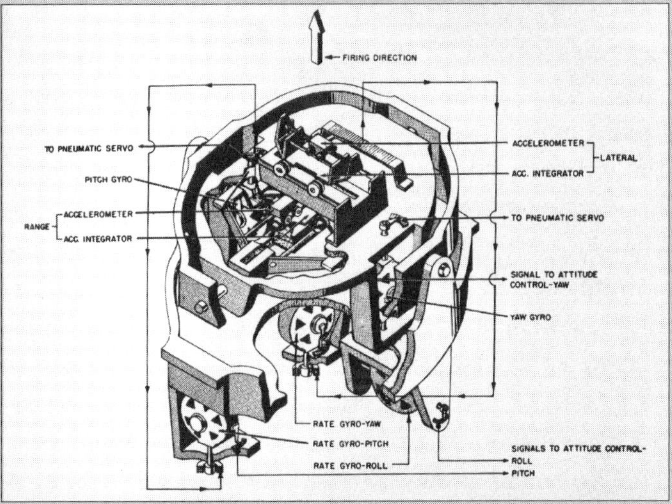

Stabilizing platform for the V-2 missile. Courtesy of National Air and Space Museum, Smithsonian Institution (SI 80-4130).

This system was designed and completed by a brilliant young engineer working under the direction of Dr. Ernst Steinhoff. (It was some time before I learned that this brilliant young engineer was Helmut Hoelzer, who would become my boss and lifetime friend.) The development of this mixing computer by Hoelzer eventually led to his invention and the building of the first electronic analog computer.

In the design of the rocket engine it became necessary to develop pumps which would inject the fuel and oxidizer into the combustion chamber at greater velocities. Many other engineering problems arose which required timely solutions. The direction and coordination of facilities scattered across wartime Germany became unworkable, leading von Braun to make the decision to bring all of the development and manufacture of the missiles in-house. This led to the building of major facilities at Peenemünde. Von Braun later followed this same philosophy at Redstone Arsenal.

Dr. von Braun came in to see Dr. Friedrich and Dr. Thiel once or twice a week. The first time he came into the office, I knew him immediately, even though I had never met him. He walked with an aggressive gait and an air of confidence, not arrogance as some have speculated, surrounded him. He was a handsome man with a good smile. Dr. Friedrich introduced me to Dr. von Braun, who gave me a firm handshake and welcomed me to the team. Then the two of them went into a technical discussion, and I got the impression Dr. von Braun used Dr. Friedrich as a sounding board and to explore some thinking on flight mechanics and aerodynamics for the space station he was then planning. Dr. von Braun would, on occasion, draw me into the conversation to determine my progress. Dr. von Braun had established the procedure of assigning some newcomers, such as me, to an experienced rocket scientist with teaching experience to establish a cadre of trained scientists and engineers.

A few times, Dr. von Braun came to consult with Dr. Friedrich at the lunch hour. Dr. Friedrich would have a sandwich from a paper bag, as we all did, since there was no cafeteria and we were miles from a restaurant. Dr. von Braun would eat macaroni from a cottage cheese jar (in those days all milk products were sold in glass containers), and at first I thought he was eating a full quart jar of cottage cheese. In a later discussion, Dr. Friedrich told me that Dr. von Braun did not eat from hunger so much as from the necessity to stay healthy. His colleagues often had to remind him when the time came to do some of the ordinary and normal things of life.

The story was told that when he reached the age of thirty-five his colleagues informed him it was time to marry. The thought of an extended courtship and engagement was not something Dr. von Braun had in his plan, so he looked to an arranged marriage. He designated a beautiful distant cousin in Germany whom he had not often seen in recent years as his target for marriage. Maria, the beautiful cousin, consented and said she had never planned on marrying anyone other than Dr. von Braun. From all accounts, the marriage was a happy one and three children came along right on schedule.

Wernher von Braun was born in Wirsitz, Germany, in the Prussian province of Posen. His parents were the Baron and Baroness Magnus von Braun. He was born in 1912; his older brother, Sigismund, in 1911; and his younger brother, Magnus, in 1919. The boys grew up in a home filled with a zest for serious reading, classical music, and good conversation in any of a half dozen languages. Their mother, Baroness Emmy, was of Swedish-German aristocracy, well educated, and an accomplished astronomer. At Wernher's confirmation, his mother gave the thirteen-year-old a telescope instead of the customary gold watch. Wernher had been experimenting with rockets. He and Sigismund had landed in trouble when they attached the biggest sky rockets they could find to a coaster wagon in an attempt to break speed records; the wagon, unfortunately, ended up in a neighbor's flower garden.

Young von Braun was not a serious student in his classes since he found so many things to explore and build outside the classroom. The turning point came when he received the book *Die Rakate zu den Planetenraumen* (*The Rocket into Planetary Space*) by Hermann Oberth. The pages in the book were a hash of mathematical formulas which made no sense to him. After Oberth had read Jules Verne's *From the Earth to the Moon*, the mathematician had written this mathematical treatise to show what would be needed for a trip to the moon. This incentive to understand overwhelmed young von Braun, and right away he transferred to the Lietz School on an island in the North Sea and was soon substituting for the math teacher.

In 1930, von Braun entered the noted Charlottesburg Institute of Technology, where he met his hero, Hermann Oberth, and graduated with a bachelor's degree in aeronautical engineering in 1932. He then went to the University of Berlin, where he received a PhD in physics in 1934 at the age of twenty-two. His PhD thesis translates into "An Analysis and

Dr. Wernher von Braun, 1962.

a Measure of the Complicated Injection, Atomization, Combustion, Dissociation, Gas Equilibrium and Expansion Phenomena That Take Place in a Rocket Motor."

During his studies at the University of Berlin, Dr. von Braun became an employee of the army charged with developing liquid rockets. In 1935, the German Luftwaffe offered five million marks as a prize to the group who could build an all-rocket, or jet, fighter plane. (Some argued that the Treaty of Versailles, which restricted the Germans from building or experimenting with powered aircraft, referred only to airplanes powered by piston-driven engines.) The army held onto Dr. von Braun and placed him in charge of the project. At the age of twenty-three, he had made it into the big time as head of the German army's effort to build a jet engine for aircraft. This would lead to his appointment as head of the German guided missile effort a couple of years later. The German army had decided that the secret weapon of the Third Reich would be the guided missile capable of bombarding England, and later, New York City.

During the Christmas holiday of 1935, Dr. von Braun discussed his new project with his mother and mentioned his need for a remote site on the sea where it would be safe to test rockets. His mother suggested the island of Usedom in the North Sea where her father had gone duck hunting when she was a child. Thus the village of Peenemünde would become world famous as the site where the first guided missile weapons in history would be launched.

The main mission at Peenemünde was to build and launch guided missiles carrying destructive explosives to harass and bring destruction to England. At the first successful launch in 1942, however, the commander at Peenemünde, Gen. Walter Dornberger, announced: "Today the space ship is born."

During the next three years, the von Braun team built and launched hundreds of the V-2s toward England. One official count states that there were 1,403 successful launches against England, with 1,115 getting through to create an "incident." The damage done by these missiles, however, was not nearly as destructive as the English had originally feared.

During the early days in Huntsville, Dr. von Braun and Dr. Friedrich usually spent the lunch hour talking computations having to do with sending a spacecraft into orbit and building a space station that would serve as a launch pad for a trip to the moon. The orbiting station looked like a big wheel with spokes and spun to simulate gravity to those on board. This concept for space flight was to show up later in the work of American film producer Walt Disney, who visited Huntsville and became a good friend of Dr. von Braun.

From time to time, a few other rocket scientists would come through for a technical discussion and an occasional bull session. They enjoyed telling stories of funny incidents or embarrassing situations in which some of them had become ensnared. It was like a group of war veterans telling the same story over and over, but each time it became a little funnier.

In one such story, set in 1945, some members of the von Braun group were flown from Germany to El Paso, and never actually entered the country legally through customs. Some months later, this was discovered by the army and had to be corrected. Those who had not come through customs were loaded onto buses and taken across the Rio Grande River into Juarez, Mexico. Before the return trip through customs, the army took them all to a restaurant and treated them to a steak. It had been some time since they had been able to order a meal as they wanted.

As the Spanish-speaking waitress went around the table, she asked how they each wanted their steak cooked. After getting through the language (English, German, and Spanish) barrier with the first order, the waitress asked the second German if he also wanted his steak *medium*. He was a big, hungry rocket scientist, who got to the floor and said, "No!" He did not want his steak medium, and as he spread his hands to the size of a small dishpan, he said "I want my steak BIG!" There were many other funny stories about situations when the language barrier was not properly cleared.

On a few occasions, Dr. von Braun and Dr. Friedrich had animated discussions about "the contract." Dr. Friedrich was very upset about it, and Dr. von Braun would make an effort to assure him that things would turn out well. After one of these discussions about the contract, Dr. Friedrich explained his concern to me. The von Braun team of several hundred had surrendered to the American forces at the end of World War II and was taken to an army base at Garmisch in Bavaria. Dr. von Braun wanted to bring his entire team to America and continue his goal of sending a rocket to the moon. After many weeks of discussions and many appeals to Washington, the U.S. Army was allowed to bring approximately 120 members of the von Braun team to America. By ship, rail, and plane they were taken to Fort Bliss, Texas, and housed at the White Sands Proving Ground where some high-altitude weather rockets had been launched in the 1930s by Goddard. Facilities there continued to be used to experiment with and improve the V-2s, which had been brought to America along with several tons of technical papers, equipment, and a good supply of spare parts.

The von Braun team was treated somewhat as prisoners of war, but the war was over, and the group did not fall into any defined category. The army was beginning to develop this group as the lone capability of the United States in the field of guided missiles. They could not be hired as civil servants because they were not citizens; therefore, a special contract and procedure was authorized so that they could be paid. The name of the procedure was called Project Paperclip—which gave some indication of its permanence. The contract was on a year-to-year basis and provided no security, no seniority, and no retirement plan.

Dr. Friedrich was in his late forties while most of the other members of the team were in their late thirties or early forties and less concerned about retirement. This issue eventually led Dr. Friedrich and a few others to leave for private industry.

Chapter 3

An Early Computation Unit

On a Friday afternoon, the personnel office asked that I report there the first thing Monday morning: a procedure used, I was certain, to give a person something to worry about and to spoil the weekend fishing.

During my training with Dr. Friedrich, I had met Les Southard, who turned out to be the best bass fisherman I would ever get to know. Les was a native of Athens, Alabama, which is about twenty-five miles west of Huntsville and near the Elk River where it flows into the Tennessee River. He knew this area from his childhood and knew where one would most likely find a large bass on a given day.

Les studied the weather report before each fishing trip as well as his fishing diary, kept from previous years. A thorough examination of rods, reels, and especially lines was made before the first cast; Southard did not count on fishing luck but rather on preparation and execution. He would tell me fishing luck is 40 percent preparation and 60 percent execution. A great football coach, Bear Bryant, who came home to Alabama sometime later, once used a similar expression about his team and I often wondered if he had been fishing with Les!

On one occasion, early in our relationship, I lost a large fish due to a frayed line which broke at a critical time in the combat with a big bass. After examining my line and seeing that it had several slightly damaged places, Les stopped the fishing and gave me a lecture. His basic point was to question why anyone with just a little sense, much less a rocket scientist, would spend several dollars and a lot of time to go fishing and risk losing the biggest fish of the day over a two-dollar line. I was somewhat surprised at how upset Southard was about my losing this good fish, but later realized this was an indication of how serious he was about catching a big bass. His point was well made and well received.

The fishing day started shortly after daylight and ended late in the afternoon. There was little time for snacking or napping; the fishing plug had to be constantly on the move, and a good bass fisherman was expected to place the plug on a spot the size of a large dinner plate at forty feet. Favorite fishing plugs used in the early 1950s had names like Lazy Ike, Cisco Kid, Bomber, Hawaiian Wiggler, and the Broken Back Shiner Minnow. For surface fishing, the favorites were the Chugger Spook, the Jitterbug, and the Lucky 13. The preparation and planning paid off for Les, and there were many days when one man could barely lift the stringer of bass he had caught.

Les Southard with a daily catch of bass, 1952.

In addition to his skill in fishing, Les was an excellent hunter and spent many happy days in the field with his favorite bird dog, Mitch. A covey rise, even in a wooded area, could see him bring down two birds. Les always kept himself trim and in good shape. He had the figure, the calmness, the demeanor, and the looks of Chester on the television series *Gunsmoke*. He even walked a little like Chester—without the limp, but with a slight starboard list.

The visit to personnel on Monday gave me the good news that I was being promoted but the bad news that I would no longer be working for Dr. Friedrich. I was shown an organization chart that Dr. von Braun had proposed and which was in the process of being approved. Following the lessons he had learned at Peenemünde, Dr. von Braun had decided there should be an in-house capability to research, plan, design, build, checkout, and launch the missiles the army was to build.

He proposed nine laboratories to be named Research Projects, Aeroballistics, Guidance and Control, Structures and Mechanics, Fabrication, Launching and Handling, Quality Control, Static Testing, and Missile Firing. The nine men named to head these laboratories were respectively, Dr. Ernst Stuhlinger, Dr. Ernst Geissler, Dr. Walter Haeussermann, Willie Mrazek, Hans Maus, Hans Hueter, Erich Neubert, Karl Heimburg, and Dr. Kurt Debus.

Later the Launching and Handling laboratory would be merged with other units of the army and the Computation Laboratory under Dr. Helmut Hoelzer would be added. In the mid-1950s Dieter Grau replaced Neubert, who joined the staff of Dr. von Braun along with Hueter. All of these men were veterans in their fields and had worked with Dr. von Braun at Peenemünde. Missing from the list was my friend and teacher Dr. Friedrich, and I assumed he had announced his intentions to leave the team.

I was to be assigned to the Flight Mechanics section in the Aeroballistics Laboratory. Dr. Rudolph Hoelker was to be the section chief, and I was to be a unit chief heading a unit called Computing and would be promoted to a GS-11 when I had completed six months of government service.

The personnel officer told me that I had originally been assigned the wrong grade (GS-10), since in civil service the odd grades above the GS-5 and below the GS-12 were for technical personnel and the even grades were for administrative personnel. It appeared that the personnel office had gone down the salary list, without regard to grade, and found the one that paid five thousand dollars annually, which had been my salary requirement. (I secretly wished I had said six thousand dollars back there in the Shanks Hotel.)

The personnel officer explained that a law, called the Whitten Amendment, required that a person be in grade at least six months before being promoted, and they hoped we could slip through the next few months without being discovered by Washington. It was good news because I would be making fifty-five hundred dollars a year when the promotion came through, and it was time to start thinking about replacing our 1937 Pontiac. At the end of the meeting with personnel, I was sent back to my original office with the Friden calculator and three empty desks.

Having spent a few weeks with Dr. Friedrich, I was in a good position to start calculating trajectories. I fully understood the differential equations which defined the path of the rocket but I did not yet know how to solve them and allow for the plotting of the path of the rocket in x, y, z coordinates. The differential equations could not be solved analytically as we had done in the differential equations courses I had taught at Tennessee Tech.

The solution and the path of the missile had to be determined by numerical or step-by-step procedures, which were not taught in college mathematics courses. There were, to our knowledge, only three books published in English which described numerical procedures to be used in solving differential equations. One, by Prof. W. E. Milne from Oregon State University became very familiar, as did a book written by Prof. J. B. Scarborough of the United States Naval Academy and one entitled *Numerical Methods* by A. D. Booth of the University of London.

My new boss, Dr. Hoelker, was skilled in numerical computations, so with his help and the books of Milne, Scarborough, and Booth, I was soon making out large computation sheets to accomplish the laborious process of computing a trajectory for the XSSM-14 missile.

This missile would have a range of some 150 to 200 miles and was designed to have accuracy less than 1 mile (as pinpoint accuracy was not yet required or possible). The XSSM-14, later named the Redstone, was to be an improved version of the German V-2 and would become the first all–inertial guided missile built by the United States.

All calculations were done in the metric system, which added another dimension of complexity for a product of the American education system. The calculations were repetitive for every time step along the trajectory of the rocket and screamed out for an automatic computer, but there were none.

In a few weeks, three new employees were assigned to our unit. Jim Golman transferred from Keesler Air Force Base in Biloxi, where he had worked for a couple of years. He had BS and MA degrees in mathematics from the University of Mississippi and was a perfect gentleman as one might expect of a man from the heart of Mississippi.

Billie Smith, a recent college graduate with a BS degree in mathematics from Florence State University in North Alabama, had taken a job with Thiokol (also located on the arsenal) but decided to transfer to the von Braun group. She was barely nineteen years old, having jumped ahead of her class a couple of times due to her basic intelligence and abilities to learn. Billie was as pretty as she was smart, and it was not long before a tall, good-looking engineering graduate of Georgia Tech started coming by at break time. In a few months, Billie took a trip to Iuka, Mississippi, with the young man and came back as Mrs. Ralph Robertson.

The third person was Joe Scollard, a GI who had been drafted and sent to Huntsville to serve his military time. Joe had a BS degree in agriculture engineering from the University of Iowa and could not wait to get back home to the family farm. He never made it; instead he married a beautiful young lady in Huntsville and remained there building missiles and a good family for the next fifty years.

Many GIs who had technical backgrounds were sent to Huntsville to do their two-year hitches and ended up, like Joe, spending their lives and careers in Huntsville. The military draft was an excellent recruiting tool for the project.

To execute the step-by-step computations, we used electromechanical calculators. There were three brands in widespread use during the 1940s and 1950s; the Friden was the most popular, followed by the Marchant and the Monroe.

The electromechanical calculator had evolved from adding machines which were first developed during the last decades of the nineteenth century.

In 1885, William Burroughs developed an adding machine with a full keyboard and a print capability; in the succeeding twenty years, Burroughs sold one million of their adding machines, becoming the dominant manufacturer.

The Standard Adding Machine Company of St. Louis simplified the keyboard and made operational improvements; Remington Rand streamlined the ten-key systems. Both of these companies sold thousands of machines.

The Victor Adding Machine Company developed a series of mechanical calculators and sold more than a million of their machines. The first part of the twentieth century saw great innovation and competition in adding machines, cash registers, and other machines to meet the needs of the businesses which were beginning to flourish.

The Friden Company was founded by Carl Friden, a Swedish immigrant who introduced a full keyboard, motor-driven calculator in 1934. The Friden did not have print capability but was rated as a high-quality, reliable calculator. The Marchant Calculating Company of Oakland, California, built a calculator based on a principle called proportional gears, where each digit of the main keyboard had its own nine-speed transmission (in automobile terms). The Marchant was built for speed and was considered to be the fastest calculator of the day.

The Monroe Calculating Machine Company of Orange, New Jersey, was formed in 1920. The Monroe model of the 1940s was very light and compact as compared with the competition, and its operation was more like that of the Friden than the Marchant.

Each of the three calculators used different mechanical processes to perform the arithmetic operations. The Friden had two keyboards, one large and one small. To multiply, one entered the multiplicand in the larger keyboard and the multiplier in the small keyboard. The Monroe was similar but much smaller physically with only one keyboard which did double duty.

The operation of the Marchant was much different. It had one keyboard in the center and a vertical column of digits on the right side. After the multiplicand was entered into the keyboard, the multiplier was entered one digit at a time. A skilled operator could keep the wheels turning continuously, as the Marchant was played like an accordion.

I chose the Marchant, Jim and Joe selected the Friden, and we all decided that Billie should select the Monroe since it was the most delicate and feminine. Each of us was biased toward the calculator we had chosen. We touted the advantages of our machines similar to those discussions around country stores in the 1930s as men argued about the merits of their Ford, Chevrolet, or Plymouth.

This bragging on our calculators led to a contest we held at least once a day. When it came close to break time, one of us would go into the hall, grab a passerby, and ask them to come into our office and write a random six-digit number on the blackboard. At the go signal, we would each start computing the square root of the number to four significant figures using our favorite scheme. It usually took two to three minutes. The last to finish bought the Cokes and the winner was given points toward some major prize to be awarded at a future time. Even though I had the fastest calculator, my colleagues found ways to beat me more often than not.

Salesmen from the firms who sold the calculators we used came by occasionally to see if we needed additional equipment. One day I heard the Friden representative coming down the hall toward our office. It was easy to recognize him because he had a distinctive gait which became more pronounced when he was carrying a forty-five-pound Friden. On this day he was carrying equipment, so I held the door and helped him in with his heavy load.

He plopped the Friden down on the corner of my desk, removed its cover, and said that this was a new model with a great new capability. He asked if we noticed anything different. None of us did, so he then pointed out an additional button with a radical sign (√) imprinted on it. He told us this Friden could calculate the square root of a number simply by entering the number in the keyboard and pressing the new button.

None of us believed it was possible and even ridiculed the idea until the salesman proceeded to demonstrate. He entered a number and pressed the button. The Friden went through several gyrations, accompanied by a squealing noise. It then produced a number which was supposed to be the square root.

We doubted its accuracy and jumped on our machines to check it out. We found it to be correct to eight digits, then asked for a second try, which was also correct.

We had just witnessed our first breakthrough in computation, and it was going to destroy our daily contest for Cokes. The price for this new system was twelve hundred dollars compared to the eight hundred we had paid for the standard Friden. This new version of the Friden replaced not only the older Fridens, but it also began to cut in on the business of Marchant and Monroe.

The computation unit, established in the aeroballistics laboratory, was charged with investigating methods to more accurately and efficiently perform the computations required, in addition to doing the actual computations. This general area was later to be called numerical analysis. Methods of approximation with names like Cauchy-Euler and Runge-Kutta became part of our daily discussions, and we concentrated on shortcuts for convergence and methods to improve accuracy. We spent so much time investigating numerical methods and approximations that we got far behind in the trajectory calculations we had been assigned.

It occurred to Jim that we could train assistants to do the calculations if they had a high school education and understood basic arithmetic. I proposed that to our boss, who readily agreed with our proposition and suggested that we might hire the wives of some of our employees. He even suggested that females might have a better skill at doing these computations than men. This was long before the feminist movement so we requested that the personnel office specifically hire women for the jobs. They did, and we soon had four ladies learning to love their calculators and wondering aloud, "What are trajectories?" They were called math aides, a designation we had selected since our classification was mathematician.

We programmed the math aides by constructing a large sheet, almost as large as the top of a desk, with rows and columns forming a large spreadsheet. The rows from left to right represented the calculations needed for one-time step of the trajectory. The heading (first row) of each column gave the formula for the calculation that was to be done to fill in the cell at the intersection of the row and column. As calculations for a row were completed, the increment in time (Δt) was applied, and

Our computation unit was responsible for supplying information to engineers that would help them design missiles to meet the requirements set down by the army. As a result of the cold war and the Korean Conflict which began in 1949–1950, the U.S. government recognized that it must again be prepared for international conflict. The United States intelligence groups learned that the Soviet Union was developing a guided missile at a time when our country had no plans for rocket development. (We had spent the war years building the ultimate weapon, the atomic bomb.) The guided missile presented a new way to deliver the big bomb as well as conventional explosives.

As our newly formed Defense Department searched its resources, it found only one engineering group capable of designing and building a guided missile, the von Braun team which already worked for the army. The U.S. Army was given the mission to develop the first United States guided missile, and quickly moved to find a proper facility for its design and construction. The move from Fort Bliss to Huntsville in late 1950 gave a new and welcome way of life for the German team, who were then allowed to bring their families together, build houses, and begin a more normal way of life.

the next row calculated. This step at the end of the row, which applied the step-by-step integration procedure, was the most difficult step to explain.

The calculations were divided into three phases called the propelled phase, free flight phase, and dive phase, since there were significant and different forces acting during each of these phases. During the propelled phase, the main force acting is the thrust of the engine; in the free flight phase the main force is the reduced force of gravity; and during the dive phase, the main force is the drag force. The time increment (Δt) was varied depending on the significant forces which were acting in the particular phase of the flight; therefore during free flight the time step could be significantly increased since external forces were at a minimum. If the total flight of the rocket was to be 720 seconds for example, there could be forty to one hundred rows of calculations; the number of rows varied depending upon the desired accuracy. The coordinates calculated at the end of each row represented a point on the trajectory of the rocket.

Three of the math aides were southerners. Valeria Wells was the wife of an engineer from Troy, Alabama; Ruby Eastwood was the wife of a physicist who had moved to Huntsville from Tennessee; and Betty MacDonald was from Athens, Alabama. Edna Flanagan, originally from Philadelphia, had worked in Washington, D.C., then moved to Alabama with her unmarried daughter, also a government employee. They had moved to Alabama since her daughter was able to get a promotion by making the move, and Mrs. Flanagan joined us.

Mrs. Flanagan could find little, if anything, about Alabama she liked. The weather was too hot; the humidity was unbearable; the people were backward; the stores had no quality goods; and you couldn't get good food as existed in Philadelphia.

There was very little air conditioning at that time and certainly none in government facilities, as it was still considered a luxury item. Mrs. Flanagan would bring her own electric fan to work and spend the day trying to get it properly adjusted. She would carry the fan home each night and hope for cooler weather the next day.

Mrs. Flanagan was senior in both years and in government service compared to the other ladies who were hired as math aides, therefore she took it upon herself to be the leader of the group. She was the one who knew all the government rules and regulations, as she had worked in Washington.

The math aides needed some books and tables related to atmospheric conditions at certain altitudes, and these tables had been classified as confidential material. The aides were told they had to be responsible for making sure their copies were in the safe in my office when they left for the day. Mrs. Flanagan interpreted this rule to mean that anytime they left the office they had to come into my office and lock up their documents. Since I was the only one who had the combination to the safe, I had to open it each time an aide needed to go to the restroom. This put the aides in the position of having to get me to open the safe and give them permission to go to the restroom. This was somewhat embarrassing for some of the ladies, and I had to override Mrs. Flanagan, who threatened to report me to security!

Mrs. Flanagan had used the Friden calculator as an adding machine on her job in Washington, and she told me that she was always given an answer to check her results. Many times when I entered the room she would be sitting with her arms folded across her ample bosom as if she was on strike, and it would be obvious she was upset. When I would ask if there was a problem, she would tell me that she had become accustomed in Washington to seeing some final results and understanding the meaning of her work, and she was not given that "down here in Alabama." She did not understand what she was doing, and it was my responsibility to explain it to her.

After a few of these incidents, I decided to give Mrs. Flanagan and the other ladies a lecture on rocketry and the calculation of trajectories.

I started by telling them about Isaac Newton, who was born in Lincolnshire, England, on Christmas Day 1642 and died in London in 1727. Newton had been born into a successful farming family and was expected to take over the farm when he reached adulthood. It didn't work out that way since Newton had a great interest in science and none in farming. At eighteen, he was allowed to enter the university at Cambridge where he became an undisciplined, unconventional, and outstanding scholar who did not follow the prescribed courses but moved ahead rapidly in areas of interest to him. His studies were interrupted by the plague sweeping through the cities of Europe, and he was forced to leave the university.

Newton continued his studies back on the farm, where one of his favorite places to read was under an apple tree. One day an apple fell and hit him on the head, or so the story goes, which led Newton to ask himself, *Why did the apple fall down instead of up, or sideways?*

This led Newton down a path to lay out the fundamental principles of the theory of gravitation and the fundamental laws which govern motion and movement of objects on the earth. The three basic laws of motion laid down by Newton can be generally stated as:

1. Unless acted upon by an external force, a body at rest stays at rest and a moving body continues moving at the same speed in the same straight line.

2. A force applied to a body gives it acceleration (*a*) proportional to the force (*F*), in the direction of the force, and inversely proportional to the mass (*m*) of the body. This can be stated as:

$$a = \frac{f}{m} \text{ or } F = m \times a$$

if constant of proportionality equals 1.

3. When body *A* exerts a force on body *B*, *B* exerts an equal and opposite force on *A*. Stated simply, "for every action there is an equal and opposite reaction."

I explained to the aides that these laws of Newton were being used to determine the equations of motion which allow us to calculate the path of a guided missile over the surface of the earth.

In addition to laying out the fundamental laws of motion, Newton developed a branch of mathematics to facilitate the calculation of the paths of moving objects which obeyed these laws. This great tool, called calculus, is two-pronged, consisting of differential calculus and integral calculus. Many physical problems can be mathematically defined using differential calculus and Newton's laws of motion, and then solved using integral calculus.

I explained to the aides, as I had done with former students in calculus classes, that we used differential calculus to set up the problem and then we used integral calculus to knock it down and solve it. We would do that for the trajectories we were calculating. From the looks on their faces, it was obvious the ladies were not very interested in Isaac Newton, except for the incident with the apple.

I approached the rocket trajectory problem by quoting Newton's third law which states that "to every action there is an equal and opposite reaction." This law is what makes a rocket go. I used the example of the backward force of a fire hose when it was opened and the example of the kick of a shotgun. I then explained that as gases were expelled from the aft end of the rocket there was an equal force generated in the forward direction which made the rocket move. The larger the force being expelled, the larger the thrust to move ahead.

The second law of Newton states that for a body in motion the force (*F*) exerted by the body is equal to the mass (*m*) of the body multiplied by the acceleration (*a*) of the body. In mathematical terms we can express this by the formula:

$$F = m \times a$$

I was beginning to see some frowns in my audience at the sight of the equal sign, but I decided to forge ahead. I explained that we could derive the differential equations which govern the flight trajectory of the missile by remembering that acceleration (*a*) is the rate of change of velocity (*v*) with respect to time (*t*) and write the formula:

$$F = m \times \frac{dv}{dt}$$

where *dv/dt* has replaced acceleration (*a*), and then change sides and write this equation as:

$$\frac{dv}{dt} = \frac{F}{m}$$

The symbol $\frac{dv}{dt}$ is a mathematical symbol (from calculus) which means the rate of change of *v* with respect to *t*, which is acceleration (*a*).

In an automobile when the driver stomps down on the accelerator the velocity increases, and the rate at which the velocity increases is called the acceleration of the auto, which is the rate of the increase in speed (velocity). In calculus $\frac{dv}{dt}$ is called the first derivative of velocity with respect to time.

We now needed to remember that velocity (speed) is the rate of change of distance (*x*) with respect to time (miles per hour, for example), and write the equation:

$$v = \frac{dx}{dt}$$

In calculus $\frac{dx}{dt}$ means the rate of change of distance (*x*) with respect to time (*t*) which is, of course, velocity. We also remembered that we can perform the same operation on both sides of the equality and still have equality.

Stated another way, if we slap one side of an equation then we must slap the other side exactly the same way if we are to maintain equality. Now we differentiate both sides of this last equation and get

$$\frac{dv}{dt} = \frac{d^2x}{dt^2} = \frac{F_X}{m}$$

or

1) $$\frac{d^2x}{dt^2} = \frac{F_X}{m}$$

The symbol $\frac{d^2x}{dt^2}$ is equal to $\frac{dv}{dt}$ and in calculus it is called the second derivative of distance (*x*) with respect to time (*t*).

F_X means the sum of those (several) forces acting parallel to the x-axis, where *x* measures the distance along the earth's surface. It should be noted that F_X will be made up of several terms and may involve all three dependent variables (*x*, *y*, and *z*).

The letter *m* represents the mass of the missile which can be approximated by $\frac{W}{g}$ (where *W* is weight of the missile in the proper units). This relationship, called a differential equation of the second order, defines the path of the missile as it moves in the *x* direction. The tough problem is, of course, calculating the several terms which make up F_X—which is what the math aides were doing as they worked their way across their wide sheets.

Now if we look at a second dimension (*z*), which is the height of the missile above the earth's surface, the differential equation which applies can be written as follows:

2) $$\frac{d^2z}{dt^2} = \frac{F_Z}{m}$$

This equation (when solved) tracks the height of the rocket as it moves along its path. The symbol $\frac{d^2z}{dt^2}$ is acceleration in the z direction and F_Z is the sum of the forces acting in the z (vertical) direction.

The third dimension is the lateral deviation from the planned flight and this is called the y dimension. A third differential equation for this dimension can be written as:

$$3)\qquad \frac{d^2y}{dt^2} = \frac{F_y}{m}$$

F_y is the sum of the forces acting in the y (lateral) direction. We call the path of the missile the trajectory, and we are calculating the path of the rocket by solving these three differential equations.

The three equations are called ordinary differential equations of the second order and must be solved simultaneously. The independent variable is t (time) and the three dependent variables are x, y, and z. Time (t) is varied in small steps from ignition to missile impact and the values for F_X, F_Z, and F_Y are computed for the succeeding time steps along the trajectory of the missile.

I related the differential equations, and the terms involved, to the rows and columns on the trajectory sheets the aides were calculating. I then explained that what we were doing was solving numerically, and simultaneously, a set of second-order differential equations.

I then asked if there were any questions. Mrs. Flanagan had one. She stated that there was always one term that puzzled her and she always wondered why we used it so much. Her question was, "What, actually, is gravity?"

I did not consider my first lecture on rocket science a great success; neither did the math aides.

Late on a Friday afternoon, the personnel office called and asked that I come to their office early on the next Monday morning to discuss a matter which needed our early attention. Les Southard and I had plans for a Saturday fishing trip to Anderson Creek, located approximately twenty miles west of Athens, so I decided to give the fishing trip my primary attention and tried not to speculate on the serious matter to be discussed with the personnel office.

On Monday, I learned that the personnel office had received a notice from Washington, D.C., that we had classified the ladies who were computing our trajectories as math aides, and there was no such civil service classification. This was a serious matter. The personnel clerk told me that we may have been doing something that could get us all in trouble. I was anxious, as were they, to clear up the matter.

They explained that it would be necessary to have a person from Washington come to Huntsville to spend about two weeks doing a desk audit and then determine the job classification, perhaps establishing a new category.

A personnel specialist from Washington arrived in Huntsville, and with one of the local personnel specialists, sat in the room with the ladies calculating trajectories. They watched every move the ladies made and took copious notes. The specialist took these notes back to Washington for analysis; we hoped for an early decision despite being told the process could take up to two months.

About three weeks after the specialists left, our personnel office called with good news: the people in Washington had completed their task. The classification for our aides had already been established several years earlier at the request of the personnel office at the Aberdeen Proving Grounds in Maryland. The engineering group at Aberdeen was charged with calculating firing tables which were used for aiming large cannons at targets under various conditions. The people calculating these firing tables were primarily ladies who used desk calculators to complete their task. They were called computers; hence our math aides would in the future be called computers.

Ironically, one of the earliest computers was Kay Mauchly, whose husband, John, led the effort to build the very first computer, the ENIAC. In later years, I once heard Kay Mauchly say she *was* a computer before her husband *built* the first one!

Many women were involved in the development of the ENIAC. Some of them were mathematicians, but a great number were simply trained to calculate trajectories and firing tables. The army research laboratory at the Aberdeen Proving Grounds had made the decision that women were more suited than men to do these calculations and at one time employed eighty women and three men.

So, as it turns out, the first computer I ever programmed was a woman! Over the years I have reflected on this, and as each new computer comes along, I compare it with my very first computer, as a man often does with his first love. No computer has ever compared favorably with the woman computer when complexity, intuitiveness, sensitivity, and tenacity have been the factors of major consideration.

Chapter 4

Dr. Helmut Hoelzer: Lifetime Friend and Advisor

Once the human computers were programmed and the Fridens were generating numbers toward establishing a new trajectory, I began to take two breaks a day in our new snack bar on the first floor of Building 111. Coffee and doughnuts were available in the morning, cold sandwiches and drinks at lunch, and coffee again in the afternoon. One day I sat down and started talking with a man who was to become my lifelong friend and mentor.

His name was Helmut Hoelzer, and he was a member of the von Braun team going back to the earliest days at Peenemünde. He had a PhD from the Technical Institute at Darmstadt, was a genius in electronics, and had many technical accomplishments in Germany.

The dissertation for his terminal degree was the description of the very first electronic analog computer, which he designed and built while working on the development of the guidance system for the V-2 missile. The computer Dr. Hoelzer built at Peenemünde was shipped to America along with a hundred or so V-2 missiles, missile parts, and documents from the V-2 development.

Dr. Hoelzer told me about his youth in Germany and how he had become a member of the von Braun group. He grew up in a village near Bad Liebenstein with parents of moderate means and rode a bicycle a few miles each day to attend the academy (high school). He received scholarships to attend college and later for graduate studies at the Technical Institute at Darmstadt. He finished his master's degree in mathematics and electrical engineering and planned to proceed to the PhD, studying under Prof. Alwin Walter at Darmstadt. However, as war spread throughout Europe, the funding for his scholarship ended. He then took a job with the Siemens Company to earn enough money to continue working toward his degree.

One evening as he sat in his quarters on the second floor of a rooming house, he heard gravel being thrown against his window. He looked out and saw his good friend Ernst Steinhoff and two other young men, one whom he had never seen before. Steinhoff asked that he join them at the nearby pub for a beer, and young Hoelzer agreed.

At the pub, the four young men drank beer, listened to music, and admired the pretty girls as Steinhoff led the discussion.

Steinhoff was a teacher and had become the local, perhaps even the world, champion in long-distance gliding, which had become a major sport in Germany after World War I. One of the most difficult tasks in gliding is to estimate ground speed on landing, and Hoelzer had used his genius in electronics to build a device which did that for himself and for Steinhoff.

The man whom Hoelzer had never seen before asked only two questions. He asked Hoelzer if it were true that he had built a device which would calculate ground speed from a glider, to which Hoelzer answered in the affirmative. A little later the young man asked Hoelzer if he thought he could build a device which could calculate speed from a flying rocket, and again Hoelzer answered in the affirmative. There was no other comment from the third man as Steinhoff continued to talk.

The visit to the pub ended and Hoelzer went back to his room. In about three weeks, Hoelzer received a civilian draft notice which ordered him to report to Peenemünde to become part of the German rocket development program. The third man at the pub had been Wernher von Braun, who was, along with

Steinhoff, interviewing Hoelzer for an assignment to Peenemünde and the German guided-missile program.

Dr. Hoelzer and I had many good talks, and I found out early on that he had an uncanny interest and concern for the welfare of his associates. He would ask me about my wife and children and how they were adjusting to life in Huntsville. It seemed incongruent to me to see a man with such technical skills so concerned with the personal welfare of others. I saw these characteristics and other unusually warm, friendly traits many times throughout our forty-five years of friendship.

Early on, Dr. Hoelzer picked up the nickname "Doc," which seemed to fit him (and not the other members of the von Braun group). Several, including Karl Heimburg, Joseph Boehm, and Karl Sendler, had a good sense of humor and horseplay like Doc Hoelzer.

In our early days, associates would ask me what Doc and I talked about. I told them that he taught me about analog computers, hysteresis loss, and Gauss integrators, while I taught him about bass fishing. Doc did not want to go bass fishing, but he was interested in the theory behind it. For instance, he would ask me what should be the attitude of the rod tip and the velocity of the bait in order for the resultant trajectory to lie on the surface of a hyperbolic paraboloid. Everything except people represented a technical situation to Doc, and he wanted to see the equations.

One day Doc and I discussed the von Braun organization plan I had seen in the personnel office, and I asked where he was going to work. He told me that Dr. von Braun had asked him to study the organization and to suggest what else might be needed to fulfill the mission they had been assigned. Doc had hoped to get the job as director of the division of Guidance and Control, where he had considerable expertise and experience. The area of developing the guidance system for the missiles to be built was his true love and where he felt he could contribute most. However, when the group had moved from

Left to right: Charles Bradshaw and Dr. Helmut Hoelzer, 1962.

Fort Bliss to Huntsville in 1950, he had been consulting on analog computers with a group at Alamogordo in New Mexico, and had been delayed for a short time. The job as director of Guidance and Control had been given to someone else, and Doc never quite got over it. There would be turf fights later which had their genesis in this von Braun decision.

One day in the fall of 1951, Doc did not show up in the lunch room for our talk. I became concerned and asked if anyone knew his whereabouts. Someone had seen him in the basement of Building 111 getting a room ready for something, so I went there immediately and found him with a broom and other cleaning materials. He told me that he was preparing the room so he could establish a computer laboratory.

When I asked, "What is a computer laboratory?" he explained that it would have electronic machinery called computers which would do what the lady computers upstairs were doing—only much faster and more accurately. I did not believe him but I respected him, so I grabbed a broom and helped him get the room in order. His laboratory soon became the tenth element of the von Braun organization.

Doc had a couple of GIs assigned to him to help start his laboratory. The drafted soldiers had come from universities in the northeast, had heard of computers, and knew something of their development. They had learned of the computer named ENIAC at the Moore School of Electrical Engineering at the University of Pennsylvania and of subsequent efforts of its developers to build commercial computers called the BINAC and the UNIVAC. These computers were electronic digital computers.

When Doc said "computer," he was referring to the type he had built in Germany, which was an analog device constructed of various electronic components and which, when properly configured and programmed, would simulate the problem to be solved. Doc intended to start his computer laboratory with the analog computer he had built at Peenemünde, which along with many of the other devices was shipped from Peenemünde. Doc's computer was in a cabinet approximately six feet wide, two feet deep, and four feet high and was a maze of components and wires which no one but Doc could understand.

The analog computer solves problems by creating an electronic model of the problem, solving it, and then transferring the result back to the original problem statement. A formal definition of the analog computer was given by G. Verroust in 1965:

> *Analog calculators* are devices in which the variables occurring in the problem to be solved are represented by continuously variable physical quantities whose values are constructed by the device so as to obey the same mathematical laws as in the problem to be solved (*The Universal History of Computing* by Georges Ifrah).

A simple example of an analog (not electronic) computer is the slide rule. In the 1930s through the 1950s a slide rule attached to the belt and carried like a sidearm in the Old West was standard equipment for the engineering student. The slide rule calculates by measuring and adding distances on an instrument with two properly positioned (sliding) sticks, and is based on the use of logarithms where multiplication is replaced by addition. The senior citizen who met logarithms along the way will recall that to multiply A by B one takes the logarithm of A and adds this to the logarithm of B to get the sum C. Then to get the result X, one takes the anti-log of C. These steps required the use of logarithm tables and an understanding of characteristic and mantissa. The slide rule replaces, becomes the analog of, the book of logarithm tables. The slide rule and the use of logarithms have disappeared since the advent of pocket calculators, and the math and engineering students of today are missing a great adventure into books of tables and the thrill of interpolation.

Logarithms were developed in 1614 by John Napier, a Scottish mathematician, to ease the problem of calculation which he detested and considered a waste of time. Napier is also credited by some with the development of the first calculating machine, historically referred to as Napier's bones. Napier's crude calculating machine was made up of ivory rods with numbers on them which could be moved and rotated to perform multiplication, but not very well.

The digital computer solves problems by performing arithmetic operations on discrete numbers and by making logical decisions based on the comparison of discrete quantities. The abacus is the earliest example of a digital computer and has been in use for approximately one thousand years. Calculation is done by moving beads (discrete numbers) strung on a wire and keeping track of the value of the beads on the different wires. Some persons were so skilled with the abacus that they were matched against some of the early 1950s automatic digital computers, often winning.

Doc unpacked his electronic analog computer and started getting it in shape to do some calculations. He gave lectures on the development of his computer and how it could be programmed to solve problems, particularly where the mathematical model was a set of ordinary differential equations.

One of the first problems Doc wanted to solve on his computer was a trajectory we were calculating through the work of the human computers. We were solving a set of second-order differential equations which would be directly suited to solution by the analog computer.

Doc had realized in the late 1930s that an electronic circuit could be built to simulate all the terms in a differential equation, and he had also developed electronic circuits for multiplication, division, square roots, and some other functions. He had been very innovative and original in the use of capacitors, coils, amplifiers, and other electronic components to perform the arithmetic functions and the processes of differentiation and integration. (An excellent, detailed description of Doc's electronic analog computer can be found in a paper by Thomas Lange in *The First Computers*, published in 2000 by the MIT press and edited by Raul Rojas and Ulf Hashagen.)

Doc suggested that I furnish a completed trajectory calculation from our human computers which would serve as a good check to determine the accuracy we could get from the analog computer. This sounded like a great idea to me, but not to my boss. When I approached him with the plan, I was told, in no uncertain terms, that my unit was responsible for doing the calculations of trajectories and I should put my full energies to that task; no results of calculations done in our unit were to be given to anyone without his approval.

The concept of protective turf began to come into focus for me, and I realized these men had worked together for the past ten to twelve years and had developed a set of personal ground rules and jealousies. I came from an academic environment where turf was less important, except for an occasional tussle between math and physics.

It became necessary for me to go underground in my work with Doc. I went to the computer lab two or three times a day, at lunch and sometimes when the workday was over, to deliver initial conditions and results from a trajectory calculation.

Things did not go well, and Doc continually made repairs and adjustments to his analog computer. The computer was almost ten years old and had been transported halfway around the world in a simple wooden crate. Doc had difficulty getting spare parts that matched; however, he was a genius at making things fit. For example, in the late 1950s he drove a Mercedes to Cape Canaveral, and a part in the ignition system went bad. There was not a Mercedes dealer within two hundred miles, so Doc went to a junk yard, spent the morning going through a mountain of junk, and found a part to get his Mercedes back on the road.

The young GIs who worked with Doc occasionally broached the subject of the digital computer which they had learned about at their university. They had heard of the company named UNIVAC which had delivered its first product to the Bureau of the Census. As contrasted with the analog computer, the UNIVAC used binary arithmetic, giving precise arithmetic results.

We had discussions with Doc and were assured by him that the digital computer had many shortcomings as compared to the analog computer. According to Doc, the digital computer was very slow, very expensive, and horrible to maintain. The basic components of the digital computer were hundreds of vacuum tubes which were very unreliable and required a great deal of electric power. In spite of many potential problems, the

John Mauchly and J. Presper Eckert, the principal developers of the ENIAC, are today considered by many to be the founders of the computer industry. Mauchly began discussing his ideas for an electronic digital computer in 1941 soon after he joined the faculty at Moore School of Electrical Engineering at the University of Pennsylvania, and he soon teamed with J. Presper Eckert, then an outstanding graduate student in electrical engineering.

During World War II, advanced weapons with greatly increased ranges required the scientists calculating exterior ballistics to do much more labor-intensive tasks to establish accurate trajectories. These calculations were done by hundreds of human computers using desk calculators. It took about three hours to calculate one very short trajectory, and there were thousands of these to do. The calculations were under the supervision of the Ballistics Research Laboratory (BRL) in Aberdeen, Maryland.

John Brainerd, director of research at the Moore School, submitted a proposal to BRL for the development of a computing device to assist with the long-range trajectory calculations. Herman Goldstine, then on active duty with the army at BRL, encouraged BRL to proceed in developing a computing device. A contract between BRL and the Moore School was executed in May 1943, calling for the development of an automatic computer to be known as the Electronic Numerical Integrator and Computer (ENIAC). The ENIAC was completed in 1945—too late for the war—and dedicated in February 1946. It was acclaimed as the first working electronic digital computer.

The ENIAC did not have the capability for internal program control; the program was entered by external switches. An extension contract was let in 1946 to correct this deficiency through the development of an improved computer called the

EDVAC (Electronic Discrete Variable Automatic Computer), which was to become the first stored-program electronic digital computer.

Dr. John von Neumann wrote a lengthy and detailed paper entitled "First Draft of a Report on the EDVAC" which is considered by many to be the fundamental paper describing the general purpose, stored-program electronic digital computer. It is not certain whether or not the computer was ever completed.

The relationship between Eckert and Mauchly and the Moore School became increasingly strained, caused primarily by the patents which had been taken out by Eckert on innovations used in the development of ENIAC. The two had clashes with von Neumann, who did not believe profit should be made by research done with government sponsorship.

Eckert and Mauchly left the university in 1946 and formed a company in Philadelphia which led to the development of the first commercially available electronic digital computer, the UNIVAC. It was delivered to the Bureau of the Census and made available for sale in 1951.

UNIVAC was later sold due to the need for additional funds. First, a company named Bally invested and then sold to the Remington Rand typewriter company, which gained control of the UNIVAC computer and moved it to Minneapolis. Eckert and Mauchly remained with the company as it moved to its new headquarters.

group suggested that I do some research to determine both the viability and availability of a digital computer.

Doc and his group continued through the first half of 1952 working with the analog computer and occasionally getting some good results. Repeatability and accuracy were two of the problems, since the weather could make a difference, as could an erratic power source. There was also the problem of getting new and updated parts for the analog computer.

We continued to have discussions about the digital computer and its development. Finally, Doc agreed to look into the purchase of a digital computer.

We located the phone number of the UNIVAC Company, and Doc made the call. The phone call was a classic and should have been recorded for the ages. The conversation between Doc and the UNIVAC representative did not occur precisely as stated below, but this abbreviated version gives the essence of the talk, which ended Doc's interest in UNIVAC for several years.

Doc started out by saying that his name was Dr. Hoelzer (he did not pronounce it the way we did) and that he was from Huntsville, Alabama. After it was determined that Huntsville was the site of the Army Guided Missile Development Division (the UNIVAC representative had never heard of either), the reason for the call came to the top of the agenda.

The UNIVAC representative asked, in a very polite manner, what he could do to help. Doc said in his very best German accent, *"Ve vant to buy a computer."*

At that time, UNIVAC had only delivered one computer (to the Census Bureau) and was working on a couple more, which some speculated would serve all the scientific community for the next ten years. After all, these computers could add thousands of numbers in a second and do the work of thousands and thousands of Fridens.

The representative told Doc that the UNIVAC computers were expensive and cost hundreds of thousands of dollars. Doc assured him we had plenty of money and that would be no problem.

Then the representative asked how we intended to use the computer and what computations we planned to make. Doc answered again in a very direct way saying, *"Ve vant to go to the moon."*

That brought the phone call to an abrupt end; this was 1952 and hardly anyone (who was rational) talked about going to the moon. The UNIVAC representative, as politely as he could, told Doc he would give our phone call serious consideration and be back with us when he determined what they could do to help. We were not optimistic.

Doc was well underway with plans to build a new and up-to-date analog computer to replace his original model. In those early days, getting approval to buy electronic parts was a problem, and we often wondered where Doc thought he could get the money to buy the UNIVAC computer. Doc assured us the German group had learned ways to steer through the administrative quagmires during the war.

He told us of the time they needed a typewriter. Unfortunately, it was during a period when procurement of all office equipment was forbidden because Hitler had decided that there was too much typing and not enough shooting. Doc's group described the typewriter as an automatic device to serve as output from a new development which would set guidance parameters on a new and powerful guided missile which could carry twice the explosive power of missiles currently under development. The unique specifications matched very closely that of a typewriter, and I could imagine Colonel Klink (a character on the television show *Hogan's Heroes*) looking over and approving the request.

Several of us still preferred the specific results one could get with arithmetic, and we started inquiring about a company called the International Business Machines Corporation (IBM).

The company had an encompassing name, and there were rumors of a computer in its future. IBM was ruled by the strong hand of Thomas Watson Sr., who had taken over in the early

IBM had its genesis with Dr. Herman Hollerith, who had developed and received patents on punched-card systems in the late 1800s. Hollerith had developed these punched-card systems for data tabulation by the Bureau of the Census in preparation for the census of 1890. It was expected that the census for that year would be beyond the scope of pencil, paper, and adding machine.

The punched card measured 7.25 inches long by 3.25 inches wide, and consisted of eighty columns and twelve rows. Data, numbers, and alpha characters were punched into the cards using a piece of equipment called a card punch.

Before the data was put into use, it would be checked by a card verifier where the punched deck was run through a verifier, and the operator would perform the same strokes (but no punches) as the card-punch operator. If there was a difference, a bell on the verifier would sound, and the two operators would have a meeting and search for the error.

After data was punched into the cards and verified, it could be read and manipulated by other pieces of equipment. A card reader printer could read the data in the cards and print the results on the printer. As the cards were fed through the card reader, an electric circuit was activated if a contact was made where a punched hole existed, and the printer would print the designated character. This piece of equipment had a large plug board which could be wired to place the character read in the position required on the printed page.

A fourth piece of equipment designed to sort the data, called a card-sorter, had twelve pockets into which one could sort the stack of cards fed through the reader. Many times multiple sorts were needed to get to the level desired. A reproducing punch allowed the reproduction of a punched card deck exactly or the shifting of data to other columns as needed. A device called a calculator-punch would accumulate data and then punch the sum into a card for further use.

Another piece of equipment called a collator was designed to merge two decks of cards into a single deck according to a prescription wired into the plug board of the device. In order to process data, a user needed all these basic components—which resulted in a room full of equipment which was named the tabulating, or tab, shop.

The equipment was generally called EAM (Electronic Accounting Machines), and many people became specialists in working with this equipment. Schools were established to train key-punch operators, mostly female, and to teach the technique of wiring the boards in the punch-card equipment. Most all of the data processing (administrative applications) done in the decades of the 1920s through the mid-1950s was done on EAM.

Business was good, and IBM was getting as many orders as it could fill for its tabulating equipment. Its main competition came from Remington Rand, which had developed its own set of EAM using ninety-column cards and round holes, probably to avoid a lawsuit with Tom Watson. Watson saw no need to spend money on developing a computer which no one was sure would be in demand by the business community and he had no interest or knowledge of what might be required of the scientific and engineering community. Above all, Watson was a good businessman who would protect the resources of the company and its stockholders. Coming along in the management of the company was Tom Watson Jr., who had a different view of what was coming in the 1950s and 1960s, and he began to make overtures about computers. UNIVAC, previously Remington Rand, had been the main competitor of IBM in building tabulating equipment during the 1930s and 1940s, and Tom Watson Jr. thought they might be on to something with computers.

1930s and built a powerful company dominant in building and servicing card-tabulating equipment.

The digital gang around Doc finally persuaded him to investigate what IBM might have underway with computers, and we located an IBM office in Birmingham which served all of central and north Alabama. We learned that the name of the branch manager was N. C. Williams, and his office was located outside the main part of Birmingham in a small business complex.

Doc and I each decided to take a day of personal leave to meet with Williams. We knew we could not get approval from the army to go on official business, and, besides, my boss was closely monitoring my relationship with Doc.

We decided to drive my nearly new 1950 solar gold Pontiac, which had replaced my very faithful 1937 model. During the two-hour trip to Birmingham, I worried about what might go wrong with our attempt to get into the digital computer business. Would Mr. Williams dodge us with another business appointment or would his company tell him to cool it with those characters thinking about going to the moon?

We arrived at the business complex and cautiously entered the IBM office, where my concern over a cool reception was soon relieved. We were met at the door by a very gracious secretary who told us Mr. Williams had been looking forward all week to the meeting. As we entered the lobby I noticed a sign that read, "WELCOME Rocket Scientists, Dr. Hoelzer and Mr. Bradshaw." *Wow! I was already being recognized as a rocket scientist.*

Mr. Williams came through his office door with a big smile and an outstretched hand. He was a friendly man who made us feel comfortable and welcome. Once inside his office, and after some coffee and sweet rolls, we got down to business. Incidentally, his secretary knew how each of us liked our coffee and which one of us preferred pecan rolls over strawberry. (In later years, I saw this attention to detail exhibited on visits to IBM plants in New York, where the IBM employees leading tours knew more about the visitors than did their next-door neighbors or the FBI.)

Doc opened the conversation by telling Mr. Williams a little about the von Braun group and how the German rocket scientists wound up in Huntsville. He then explained that America needed to add guided missiles to its arsenal of weapons to protect the country from other nations which might try to do harm. (Doc and other members of the von Braun team were sensitive to the perception that they might be building weapons for aggressive or expansion purposes.) Mr. Williams knew, without asking, to what country Doc was referring.

Doc explained that the computer was to be used for solving a system of differential equations which defined the flight of a guided missile and to control the attitude and path of the missile in flight.

Doc's unique way of describing the problem was to hold a pencil between his thumb and forefinger and describe the motion in flight. With the pencil held aloft he would demonstrate three degrees of freedom in a missile flight: missile roll, pitch, and yaw. Doc would roll the pencil to demonstrate roll, then move it up and down to demonstrate pitch, and finally, from side to side to demonstrate yaw. Doc would always end the description with an emphatic *yaw, yah* and sometimes he would get so animated he would end the sentence with *yah, yaw.*

After the demonstration, Doc started describing the differential equations of motion, including those which define the control system of the missile.

Mr. Williams was doing the best he could to follow the lecture, and as he leaned across his desk to concentrate more closely, you could almost see the glaze coming over his eyes.

I felt certain he had concentrated on accounting in college and had never been inside a calculus book. However, Mr. Williams was a trouper and tried to think of questions he could ask, and I wanted to help him.

During the course of the meeting, it became clear to me that IBM did not yet have a computer for sale, but Mr. Williams *said* they did, and Doc ordered two of them! As we left the IBM office, I was more or less stunned.

I reflected on what I had witnessed during this meeting many times in the coming years. That first encounter explained to me how IBM was able to dominate the computer industry in just the next six years. They sold themselves and futures.

We returned to Huntsville to await the computers Doc had ordered. In the meanwhile, things were getting busy in my unit, and I was unable to spend much time with Doc and others in his laboratory. Computations were required to be more and more accurate as the date for the first launching of our country's first guided missile was nearing. The XSSM-14 we had worked on for a year was named the Redstone missile, and was to have a range of 150 to 200 miles and a planned accuracy of less than a few hundred feet.

I had been given another job to do with respect to the first launch of the Redstone, and this new task gave me some concern. At the launch, I was to be at the launch site to calculate the effect of the existing winds at the time the control program would be initiated and to calculate the additional effect winds might have on the subsequent trajectory.

This would involve receiving the wind velocities and direction from the air force, which would be collecting the data from balloons aloft. These would be provided in feet per second for velocity and in degrees for direction.

It would be my job to very quickly calculate the resultant wind forces (in meters per second) parallel to and perpendicular to the flight path (in the *x-y* plane), at various altitudes. The engineers on site would use my calculations to make required adjustments to the control program which had been designed and built by the Guidance and Control Laboratory.

In late 1952, as we continued to plan for the first Redstone launch, a truck pulled up to Doc's computer laboratory. In the passenger side of the front seat was a man dressed in a neat blue suit with a conservative red tie. His name was Chuck Henderson, and he was the customer engineer who came with the IBM computer.

The new computer was moved into the prepared room, and the shades were pulled so that no one could see the inside of the computer when the cover was removed by the engineer. Everything was very hush-hush.

Finally, the big day came; we were allowed to go in and look at our computer. We soon learned it was not the electronic computer we were expecting, and that it was not ours. IBM did not sell computers; they only allowed you to lease their computer for a few hours a day under the watchful eyes of the customer engineer and the sales representative.

The machine was a card-programmed calculator (CPC), consisting of four pieces of tabulating (electronic accounting

The CPC consisted of a 402 accounting machine, a 604 electronic calculating unit, a 521 gang summary punch, and an auxiliary storage device numbered 941. This latter-named unit gave a storage capability of 480 decimal digits. The principal function of the 402 was to read cards and print. The 604 was used to add numbers and punch the result into cards for printing on the 402 or for use in future calculations.

All computations on computers are done with the four arithmetic operations of add, subtract, multiply, and divide. The comparison operator—deciding whether or not one number is larger than the other—also becomes essential. With the ability to do these five things, one can program the world (given enough time).

The only one of these operations immediately obvious on the CPC was the addition function in the 604. The 604 would handle negative numbers, so that took care of subtraction. A series of additions took care of multiplication, and some clever arithmetic would solve the division problem. The comparison problem could be taken care of through addition, the use of negative numbers, and looking at the result to determine if it was positive or negative.

Someone had to tie this together in a structured manner which would allow programming. This had been investigated by some persons at United Aircraft and North American Aircraft who experimented with connecting the 407 and the 604. They encouraged IBM to consider marketing the system as the card-programmed calculator. The leader of the group was an engineer named Ramshaw at United Aircraft, and through IBM, we were able to learn what they had developed so we could begin to write programs to calculate trajectories on the CPC.

Each instruction (add two numbers, for example), was on a single card, thus the name card-programmed calculator. The instruction was executed and then another card was read. The entire computer program was a deck of cards.

If iteration was required, the deck would be read over and over until the convergence criteria were met. To perform iteration, the programmer would generally make several copies of the program, producing a large deck, then run the deck until the convergence was accomplished.

In the case of trajectory calculations, each line (time step) required the same arithmetic. If the trajectory required 120 time steps, the deck would be reproduced (120 copies with time adjustments) and run through the card reader.

The time required by the CPC to go through one line of the trajectory code might be three or four minutes, hence a trajectory of 120 time steps would require approximately six to eight hours. This represented a huge improvement in time over the human computers who took approximately one week to complete an average-length trajectory.

machine) equipment cabled together to allow some calculations to be done.

My job was, once again, to smuggle initial conditions and results to the Computer Center so the programs could be checked.

The next necessary move was to obtain approval to do some *official* trajectory calculations. It was decided that Doc would do a selling job on the new computers. When Doc set about to sell, there was none better. (Sell Doc did—even though he was still convinced that analog computers were superior to the digital computer, and that this superiority would soon be recognized. His concern with the digital computer was still with the electron tube and its mean time to failure.) There were fewer electron tubes in the CPC, which, as a calculator, did not qualify as an electronic computer, and the CPC proved to be reliable and rugged since it was a combination of tabulating equipment which had been in use for many years.

Doc was getting some jobs to program and to run on the CPC. The first CPC programmers were army privates who had been drafted (the Korean War and the cold war were well underway) and were serving their army hitch in Huntsville.

As programs were checked out and placed in production, the need for another type of employee was evident. It was determined that persons who understood how to wire the boards on the EAM equipment and how to arrange card decks for production were needed. They would also feed cards, sometimes all night long, into the CPC and make adjustments to the program deck as directed by the programmer.

These persons would be called computer operators. It appeared the best source would be persons in EAM operations. Doc was able to hire good operators, some who spent the rest of their careers working for him.

Charlie Cozelos had experience with EAM in the military and had a job as a cab driver on the arsenal when Doc found him. Lou Taylor was a career EAM person and was recruited from another installation. Charlie King was a bellhop in the Russell Erskine Hotel, and because of his energies and quick thinking became an excellent operator. A little later, Jerry Dean transferred from another group and also became an excellent operator.

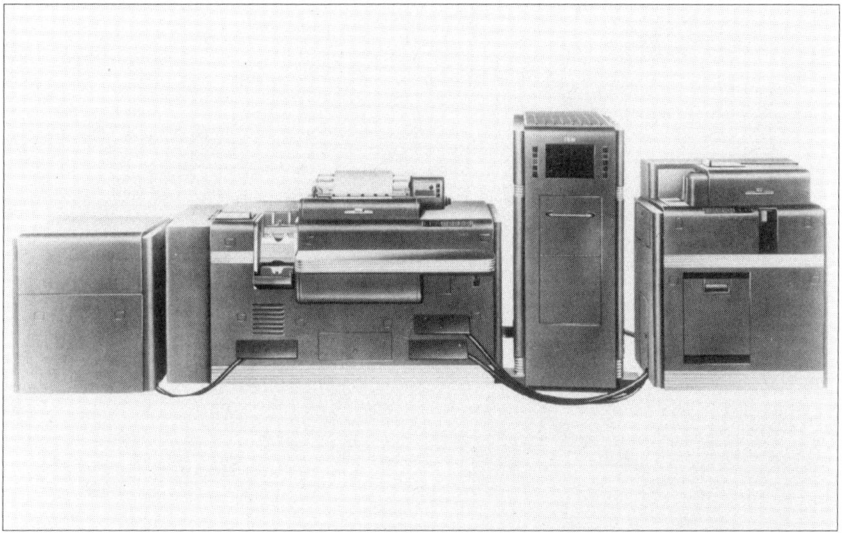

IBM card-programmed electronic calculator (CPC). Courtesy of IBM Corporate Archives.

The CPC was soon overloaded and run on extra shifts. The second CPC was delivered, and it was conjectured in those early days that computing would follow Boyle's law of gases. That is, the workload would expand to fill the resources, regardless of the size of the resource. IBM made improvements to the CPC and announced a model II which had a little more memory and some better techniques for preparing and running programs.

The CPCs served as a good transition from electronic accounting machines to the computer and also served to satisfy the IBM customers while awaiting a computer. Many technical and engineering concerns used the CPC as their first computer and accomplished many important calculations. The CPCs introduced the concept of computer programming to many and developed some early computer programmers who spent their lives working with computers.

Chapter 5

First Launch of the Redstone Missile

In the early summer of 1953, the first launch of the Redstone missile was scheduled for mid-August. This gave me a couple of months to determine how to make the calculations of resultant wind forces and their possible effect on the flight of the missile. The calculations would have to be made quickly for several altitudes and would require something other than a standard slide rule. Calculating resultant forces requires the use of trigonometry (vector analysis), while the conversion in dimensions is arithmetic.

There would not be time to use tables to look up the trig functions, so I designed and built a circular table look-up containing the trig functions. The homemade gadget was an analog computer built with an inner circle which rotated, and an arm levered from the center which could go the full 360 degrees. This circular device was divided into sections and calibrated to perform the trigonometry I would need. A slide rule would do the arithmetic.

Traveling from Huntsville to Cocoa Beach was an all-day trip on a Martin 404 from Huntsville to Atlanta and then on to Orlando. The Martin 404 flown by Eastern Airlines had two propeller engines and a capacity of perhaps thirty-five people. There was no public airport for commercial flights into and out of Orlando; the airfields at McCoy Air Force Base located approximately twenty-five miles from the city were used.

If we arrived in Orlando in the early evening, we would spend the night. There was one, and only one, sizable modern hotel (the Cherry Plaza) and we all wanted to stay there. Sometimes we had to stay in one of the motels along the highways running through town, and these motels were not nearly the same quality. There were only a few places to eat in Orlando. One favorite place was named the Duck Inn, located along one of the highways running through Orlando. It was small family place with great seafood, including lobster for about three dollars and a regular meal for about one dollar.

The army provided small buses for the trip to Cocoa Beach, but if you were lucky and traveled in a group, a rental car was sometimes allowed. The trip over the small two-lane highway was some sixty to seventy miles and took a couple of hours. There was only one motel on Cocoa Beach, named the Silver Sands, and one good restaurant on the strand named Bernard's Surf Room. I ate my very first heart of palm at Bernard's and also my first Caesar salad made with a raw egg. (My dad never believed I had started eating palm trees and once told me that we ate a lot of things during the Depression including rabbits, squirrels, and an occasional possum, but never a tree!) The Silver Sands Motel filled up early, and many of us had to stay in the nearby town of Cocoa at one of the two or three old small hotels and make arrangements to carpool out to the cape.

The U.S. Air Force had used the site on Cape Canaveral to launch some drones and small rockets over the Atlantic, but nothing the size of the Redstone missile. A rocket called the SNARK had been fired from the site by the air force; the firings had not been a booming success as several went into the sea not far from shore. A warning was given to all of us as we left Huntsville: be careful of the SNARK-infested waters off Cape Canaveral!

There was one small blockhouse, so named I suppose, because it was built out of cinder block. The launch site was a simple concrete pad with an erector crane and cabling which ran along the top of the pad back to the blockhouse. A few of the launch crew were assigned duties in the blockhouse; the rest of us had to find our work stations in the area near the launch site. I chose my workstation to be at the base of an old crane which sat about fifty yards from the launch pad. I carried along a small sign which said "Computing Office," so I could be found by the couriers bringing information on wind velocities and direction. I picked a flat spot on the base of the crane where I noticed there was crawl space under the crane big enough to shelter a man—just in case.

The unit operating out of Patrick Air Force Base, just down the road from the cape, was to send balloons aloft on the day of the launch, starting roughly four hours before the launch time. Wind direction and velocities would be transmitted from the balloons to the ground site where they would be recorded and sent by courier to my computing office. After I had completed the calculations, a courier from the Guidance and Control Laboratory would pick them up and deliver them to the person responsible for setting the control program on the first Redstone.

Fueling the missile began approximately two hours before the launch. The fuel was alcohol and the oxidizer was liquid oxygen (lox). The alcohol would be loaded first and the lox second because lox has a fast evaporation rate.

At ignition, the fuel and the oxidizer would be pumped into the combustion chamber under pressure and then ignited. The gases generated by the explosion would be concentrated and then released through the finely tuned exhaust nozzle to gain maximum thrust in a direction exactly parallel to the vertical axis of the missile.

The Redstone's all-inertial guidance system used a gyroscopically stabilized platform and was guided during the propelled phase by tail fins with movable rudders and a movable refractory carbon vane mounted in the rocket exhaust. The missile would move vertically upward until changed by a deflection of the guidance vane to direct the missile down the Atlantic Missile Range according to the design of the alpha program.

On the day of the launch, August 20, 1953, the winds were reasonably calm, a fact which gave me great relief. As it turned out, the winds were the least of the worries of Dr. von Braun and his staff. They were concerned about the performance of the engines, the stability of the missile, an explosion on the pad, or some other catastrophic event. There was, however, some quiet confidence that all would work well since these engineers had a lot of invaluable experience from their days at Peenemünde. As I was to learn throughout my twenty years of association with the von Braun team, no detail was small enough to be ignored or left to chance.

The missile was fueled and ready to go, and the countdown was nearing completion. I had done my calculations, and there appeared to be no need to adjust the control program. Liquid oxygen was escaping in a white plume as I climbed higher on the crane arm to see the exhaust from the engine at ignition. All the while, during the climb up the crane, I kept one eye on the hole in the base plate since I was not as cool or confident as my German colleagues.

I could see Dr. Ernst Geissler, head of the Aeroballistics Laboratory, and one or two others gathered behind a large metal shield with a cutout covered with glass within just a few yards of the base of the missile. They wanted to see initial ignition, study the jet stream in the first one or two seconds, check for the stability at liftoff, and carefully observe any anomalies which might occur. They had done this in Peenemünde, but the cautious United States Army was to forbid this practice in the future.

Ignition occurred. The engine gave its mighty roar, and for a second or so, nothing happened, then the missile slowly rose from the pad straight and true. In a couple of seconds, the crane arm on which I was perched felt the vibration, and I heard the solid roar from the rocket motor. As the missile gained altitude, the plume from the rocket motor became longer, and the vibration

A Redstone missile at launch. The author performed the calculations for the first launch occurring August 20, 1953.

continued like a gigantic rattle from a stream of explosions. The Redstone was flying perfectly!

The next critical question to be answered was: Would the control program work and take the Redstone down the path which was later to become known as the Atlantic Missile Range?

The Redstone climbed to a height of about four miles and headed over the Atlantic when an anomaly raised doubts about its path; consequently, the missile was destroyed with a signal from the block house. Later there was considerable discussion as to whether the destruction should have taken place since instrumentation to detect anomalies was still very crude. Nevertheless, the launch team had the safety of the civilian population foremost in their minds.

The main objectives of the test had been successful since the engines performed well, the launch stability was good, and the missile held together well. It was necessary to refine the control program, but this was a continuing process through all the missile programs in which we would engage.

I had decided that, as soon as the Redstone was launched, I would ask my boss for a transfer to the Computation Laboratory under Dr. Hoelzer. It had become obvious to me that the future of computation was with machinery and not with the wonderful ladies performing the task at that time.

I was not quite prepared for the resistance I would get. I was told that the Aeroballistics Laboratory had trained me for two and one-half years, and I was more important to them than to Dr. Hoelzer. The Germans had come from a wartime government which had a civilian draft that placed people where they would best help the cause. My boss would not immediately release me for the transfer. (I was later assured that an appeal to Dr. von Braun would have been successful.)

I decided to get into computing by looking for another job. There were only three places in the Southeast with computers installed or planned, and I visited them all.

The Oak Ridge National Laboratory (ORNL) had announced that it would soon be installing the newest computer descendant from the von Neumann line, making it the fastest computer in the world. I went there first since I knew Dr. Alston Householder, head of the mathematics panel, from my days as a graduate student at the University of Tennessee.

Security was so tight at ORNL that Dr. Householder met me at the guard building outside of the secured area. There was an opening I might fill, so Householder gave me a stack of papers to complete. One form was the personnel application. The other form, which was much longer and more detailed, was the form to request security clearance. As we parted, Householder said he hoped to receive them back from me soon.

The next stop on my job odyssey was the Lockheed plant in Marietta, Georgia. As I passed through Chattanooga, my brother-in-law Russell Scott joined me for the trip to Eglin Air Force Base. He was a veteran of the Army Air Corps and wanted to visit the facilities at Eglin and perhaps see one of the jets which were then being introduced into our military. He had heard of new planes under development, and we thought there might be a chance to see one of them in flight.

For the interview at Lockheed I was escorted to the computer group on the mezzanine of a very large building. I later learned that all aircraft plants looked the same and had the same mode of operation. It was much different from the campus-like atmosphere which existed on Squirrel Hill.

Bob Bosak and Jack Reynolds, who had moved from the California plant of Lockheed to head the new group in Marietta, interviewed me and described the work being done at the Marietta plant. Lockheed had installed two CPCs, with associated tabulating equipment, and used them for certain applications in structures. I was familiar with all of their equipment. After a short interview, I was encouraged to complete the applications and return them as soon as possible.

My brother-in-law and I left Marietta just after noon and began to travel the long and winding roads down to the panhandle of Florida. Early in the evening, we arrived in DeFuniak Springs and decided to spend the night. There was only one place to stay; it was an old and nearly abandoned wooden hotel which sat by a large lake. (It had probably been a wonderful resort hotel during the 1920s.) The hotel had no restaurant, and since there was not one open in the town, Russell and I ate supper at a local gasoline station which sold us some bread, cheese, and soft drinks.

Eglin Air Force Base was, and still is, one of the most expansive military bases in the world. There were three commands at the base performing different functions for the air force. The computer group I had been in contact with was within the group called the Armaments Testing Command. The computer group was headed by Ken Woolsey, who was in civil service, and by Fred Davis, who worked for a company named Vitro, a support contractor for the Armaments Center. I was interviewed by these two men who were to remain my friends and colleagues for the rest of my career.

The group at Eglin was using some tab equipment to do computing, and they were installing a CPC. After a thorough interview, I was again provided the application forms to complete and assured there was a good chance I would receive an offer from Vitro.

Within a few days, I received job offers from the three installations I had visited. This was not a great compliment because, at that time, I was probably one of less than two thousand people in the world who knew anything about computers. All three offered a salary of about one thousand dollars a year more than I was making.

My first choice was to go to Oak Ridge, but there was a hold up; in those days of Senator Joe McCarthy and his hunt for communists, security at nuclear sites was extreme. The letter from Oak Ridge said I would have to have a Q clearance before employment could take place, and that could take three to four months. I could have gone to work immediately at Lockheed or Eglin, but I decided to wait and go to Oak Ridge.

I stayed in touch with Dr. Householder at Oak Ridge, and finally my Q clearance came through. These clearances were not easy to get, and even though I had been an officer in the navy with top security clearance, I had to be checked back to my childhood in my home town. They wanted to make sure, for example, that my dad hadn't belonged to a labor union which had some ties to communism. No one wanted to come under the attack of Senator Joe McCarthy.

Chapter 6

What Came Before: History of the Atomic Bomb

In the 1890s, scientists in Europe began observing ores which were giving off a mysterious energy. These ores with yellow streaks were brought back from mines in the Congo and in Czechoslovakia, and showed up in the research laboratories of Paris, Berlin, and other cities in Europe.

Scientists noticed that these pieces of ore did not change their size, shape, or mass, even after months of being observed giving off a continuous stream of energy. Some scientists began to speculate that the atoms of the material were giving off part of themselves in the rays which were being emitted. This was in conflict with the thinking of the scientific community which for hundreds of years had assumed that atoms which made up the universe were single entities. Atoms were thought to be like little steel balls which fit together like ball bearings to form an object.

A French scientist named Henri Becquerel was one of the first scientists to observe that the pitchblende ore, which contained uranium, fogged up his photographic plates even when they were covered. Madam Marie Curie was another scientist to become interested in these strange rocks, and in 1898 coined the word "radioactivity" to describe the process which she observed happening with the pieces of ore she carried around in her shirt pockets. These pieces of ore were emitting trillions of particles without changing their size or shape. Curie did not understand how these rays could put out so much energy from such a minute amount of mass.

Einstein had not yet postulated his fundamental law which would have helped Madam Curie calculate what was happening. Einstein stated that the energy created when mass is changed can be calculated by the formula $E=mc^2$. This little formula, establishing the relationship between energy and mass, states that an extremely small mass when transformed into energy would equal the small mass multiplied by the square of *c*, the speed of light. The speed of light is 670 million miles an hour and when squared equals 448,900,000,000,000,000 miles per hour.

This large number would still be a significant number when multiplied by an infinitesimal small number. The radioactive dust from the ore Madam Curie carried around in her pockets spewed off rays of extremely high velocity, penetrating the bones of her body, and perhaps giving her leukemia, which caused her early death.

As the twentieth century opened, the accepted scientific thinking was still that the atom was a solid element and the fundamental building block for the universe. One scientist who had some doubts, and who became the renegade, was Ernest Rutherford. He was a farm boy from New Zealand who showed brilliance at an early age and was given scholarships in New Zealand and then England, where he received his doctorate from Oxford. He spent a few years at McGill University in Canada, where he became an honored professor and then moved to Manchester University in England.

By a systematic study of the rays emanating from the lumps of pitchblende ore, Rutherford learned that there were two types of radioactivity. One was easily absorbed and Rutherford named this an alpha ray. The other was a more penetrating type and was named a beta ray.

Rutherford and his colleagues studied these rays and determined they had peculiar characteristics. When certain rays hit an obstruction, they would bend. Rutherford soon decided that these were not simply rays emanating from the atom, but they were small charged particles that were really part of the atom itself. This placed Rutherford in further conflict with scientific thinking which had been in place for centuries; the word "atom," in fact, comes from the ancient Greek word meaning uncuttable.

Rutherford and his colleagues continued to make basic discoveries about the atom by fully exploring the particles coming from the atom. Rutherford came to the postulation, which turned out to be correct, that the atom had a nucleus made up of positively charged particles (protons) surrounded by a vacuum in which negatively charged particles (electrons) moved in given orbits. There might be other particles in the nucleus, later determined to be neutral particles which were named neutrons. Rutherford theorized that the nucleus was very small but made up more than 95 percent of the mass of the atom.

Perhaps the most brilliant of the colleagues to join Rutherford in his research effort was Niels Bohr from Denmark. Rutherford and Bohr were a perfect match because Rutherford was primarily an experimental physicist while Bohr was more inclined toward the theoretical.

During his tenure with Rutherford, Bohr became acquainted with George de Hevesy, a scientist from Hungary, and learned about radiochemistry. With this knowledge, Bohr set about to rough out what he called the radioactive displacement law and he continued working on the completion of the periodic table of the elements. The periodic table, the large chart seen in almost all college and high school physics classrooms, describes the ninety kinds of atoms which occur naturally on the earth. This table of elements, when properly interpreted, tells the number of protons and neutrons in the nucleus and the number of electrons in orbit around the nucleus.

Bohr later returned to Copenhagen and established the Niels Bohr Institute for Theoretical Physics to study quantum physics (a branch of physics devoted to the study of atomic, nuclear, and solid-state physics).

The Rutherford theory of the atom launched a worldwide effort to split the atom: to go inside to determine if the theories were correct. James Chadwick, a colleague of Rutherford, took the first step when he discovered the neutron in 1932. (Rutherford was sure the neutron was there and had urged Chadwick to pursue it to conclusion.) The neutron was so named because it was neutral, possessing neither a positive charge nor a negative charge. It might be able to penetrate the nucleus of the atom without being attracted to either the proton or the electron. This discovery gave new hope and a renewed impetus to unlock the secrets of the atom by bombarding the nucleus with the neutron.

Otto Hahn, with a doctorate from the University of Marburg in Germany, joined Rutherford at McGill University in Montreal, Canada. While there, Hahn became challenged by Rutherford to see what he could do with the atom, and Hahn became one of an amazing nine associates of Rutherford to be awarded a Nobel Prize. No other person challenged, motivated, and created admiration as did Rutherford, and no one else saw as many of his associates become Nobel laureates.

Otto Hahn returned to Germany and the University of Berlin from Canada, and found an ideal collaborator in Lise Meitner, who was the second woman ever to be awarded the doctorate from Vienna. Meitner had come to Berlin to attend Max Planck lectures in theoretical physics and stayed as a collaborator with Hahn. Meitner and Hahn worked together for many years at the Kaiser Wilhelm Gesellschaft, where special arrangements had to be made for Meitner since she was the only woman allowed in the facilities. They both left to serve Germany during World War I, and after the war, returned to their jobs and continued in a program to bombard uranium with neutrons and identify the products produced.

As Adolf Hitler came to power, it became obvious that Meitner (who was part Jewish) would have to leave Germany. She found a comfortable place in the Nobel Institute in Stockholm, not too far from the Niels Bohr Institute in Copenhagen, where her nephew, Otto Frisch, worked.

The two of them (Meitner and Frisch) formed a good team, and they were able to continue communication with Hahn and an old colleague, Fritz Strassman, in Berlin. Strassman and Hahn continued to bombard atoms of uranium with neutrons and sent their results to Meitner, who was more skilled than anyone in identifying the atomic makeup of the elements produced.

In one experiment of bombarding the uranium, Hahn had produced barium, which is a much lighter element than uranium. In evaluating the barium, Meitner and her nephew discovered that Hahn and Strassman had, in fact, split the atom.

Hahn did not know the split had occurred until Meitner and her nephew evaluated the results of the split atom. Instead of splitting, Meitner reasoned that fission (explosion) had taken place when another neutron was crowded into the already-full nucleus and nuclear fission was the result. This discovery was made in 1938.

News of the splitting of the atom and its enormous possibilities spread throughout the world. With the splitting of the atom and its release of neutrons, which could themselves be splitting other atoms, the possibility of a chain reaction was considered. Could a chain reaction be sustained? Could it be controlled? Should a chain reaction even be tried?

A nuclear chain reaction did, in fact, take place in 1942 during the early days of World War II. It was planned and executed by Enrico Fermi from Italy, then at the University of Chicago.

SCIENTIFIC EXODUS FROM CENTRAL EUROPE

During the early 1930s many scientists and university professors fled Germany and central Europe due to the persecution begun under Adolf Hitler of persons of Jewish ancestry. Many of them went to the United States, including Albert Einstein from Germany; Enrico Fermi from Italy; and John von Neumann, Edward Teller, Eugene Paul Wigner, and Leó Szilárd from Hungary. These latter four, along with Theodor von Karman, George de Hevesy, and Michael Polyani, all about the same age and from the area around Budapest, made up about half the scientific geniuses of the early twentieth century. They were sometimes called the "Men from Mars" since no one could explain how such a small area of the world could produce such a large class of world geniuses.

Einstein had moved to a position at the newly created Institute for Advanced Studies at Princeton. He had followed the work of Lise Meitner, Leó Szilárd, Otto Hahn, Enrico Fermi, Eugene Wigner, and Werner Heisenberg and knew full well the significance of the split atom, which could lead to a possible nuclear chain reaction and its implications relative to a nuclear explosive device.

Einstein was a pacifist and did not want to get involved in the war effort in any way. His only motivation to take part in encouraging the United States to develop a nuclear weapon was his fear that Hitler might also have access to this awful force. His participation in encouraging the United States to become involved was greatly promoted and engineered by Leó Szilárd and Eugene Wigner.

In looking for a way to gain an audience with American President Franklin Roosevelt, Wigner and Szilárd were able to make contact with Alexander Sachs, who was a noted economist and an intimate of the president. Sachs was interested in the proposal Szilárd was making and agreed to deliver a letter from Einstein directly to the president. The letter was signed by Einstein, who later stated that it represented one of the big mistakes of his life. The famous letter of August 2, 1939, follows:

> Sir:
>
> Some recent work by E. Fermi and L. Szilard, which has been communicated to me in manuscript, leads me to expect that the element uranium may be turned into a new and important source of energy in the immediate future. Certain aspects of the situation which has arisen seem to call for watchfulness and, if necessary, quick action on the part of the Administration. I believe therefore that it is my duty to bring to your attention the following facts and recommendations:

In the course of the last four months it has been made probable—through the work of Joliet in France as well as Fermi and Szilard in America—that it may become possible to set up a nuclear chain reaction in a large mass of uranium, by which vast amounts of power and large quantities of new radium-like elements would be generated. Now it appears almost certain that this could be achieved in the immediate future.

This new phenomenon would also lead to the construction of bombs, and it is conceivable—though much less certain—that extremely powerful bombs of a new type may thus be constructed. A single bomb of this type, carried by boat and exploded in a port, might very well destroy the whole port together with some of the surrounding territory. However, such bombs might very well prove to be too heavy for transportation by air.

The United States has only very poor ores of uranium in moderate quantities. There is some good ore in Canada and the former Czechoslovakia, while the most important source of uranium is Belgian Congo.

In view of this situation you may think it desirable to have some permanent contact maintained between the Administration and the group of physicists working on chain reactions in America. One possible way of achieving this might be for you to entrust with this task a person who has your confidence and who could perhaps serve in an unofficial capacity. His task might comprise the following:

a) to approach Government Departments, keep them informed of the further development, and put forward recommendations for Government action, giving particular attention to the problem of securing a supply of uranium ore for the United States;

b) to speed up the experimental work, which is at present being carried on within the limits of the budgets of University laboratories, by providing funds, if such funds be required, through his contacts with private persons who are willing to make contributions for this cause, and perhaps also by obtaining the co-operation of industrial laboratories which have the necessary equipment.

> I understand that Germany has actually stopped the sale of uranium from the Czechoslovakian mines which she has taken over. That she should have taken such early action might perhaps be understood on the ground that the son of the German Under-Secretary of State, von Weizsäcker, is attached to the Kaiser-Wilhelm-Institut in Berlin where some of the American work on uranium is now being repeated.
>
> Yours very truly,
> Albert Einstein

In the meeting with the president where this letter was discussed, Sachs reported that the president was very interested and stated that the matter must be followed. He appointed a board headed by Dr. Lyman Briggs, director of the United States Bureau of Standards. This board had its first meeting two weeks after the meeting with the president and was attended by Wigner, Teller, and Szilárd—but not Einstein.

In 1940, the British received information which caused them to be concerned that a practical atomic bomb could be built. They rushed a secret memo to Washington, which ended up in the hands of Lyman Briggs. Some felt that Briggs, who had been a government employee since 1897, might not be aggressive enough in following through with this project. The Briggs Commission, however, did encourage and gave some funds to universities to explore the information, describing the possibility of chain reactions and their implications in the field of nuclear weapons.

Research moved ahead at a more or less academic pace until the United States was in World War II and the Manhattan Project was later formed in 1942. In December 1941, the United States entered the war after being attacked by the Japanese at Pearl Harbor. This was a war for survival of the free world, and there was great concern that a nuclear weapon could turn the tide.

Leó Szilárd cannot be given enough credit for his efforts to get research in nuclear fission underway in the United States. This tenacious Hungarian immigrant was never deterred in his efforts to get his message to the highest levels.

During the 1920s, a strong, personal bond formed among the scientists working on the uranium problem. It began at Göttingen, where many of the scientists spent some time during summers and sabbaticals. This professional interaction led to personal relationships which would transcend politics and war. Niels Bohr from Copenhagen was the most famous of the group and became its unofficial leader.

When the war began, it appeared many of the scientists in Germany with Jewish ancestry were in danger. Bohr sent them all private letters inviting them to escape to Copenhagen, where he would provide bed, board, and a laboratory. Many of them accepted his invitation and were met at the train station in Copenhagen by Bohr or his representative.

In 1943, the German army was overrunning Denmark and threatening the laboratories of Bohr. He managed to escape to England through Sweden, and proceeded to alert the Allies to the danger of the nuclear devices Hitler may have had underway. Bohr became an important advisor to the American effort to develop a nuclear weapon, and was taken directly to Los Alamos, New Mexico, when he arrived in the United States. Bohr, like so many other scientists, provided this assistance because of his concern that the effort in Germany might succeed and allow Hitler to extend his dictatorial power throughout the world.

After many months of prodding and continued concern by American and British scientists that Germany was well underway with nuclear research which could lead to a very destructive weapon, the United States initiated research efforts. These efforts got underway at universities, particularly at the University of Chicago and the University of California at Berkeley.

In early 1945, when the war in Europe had been won and Hitler was dead, Szilárd and other nuclear scientists went to great efforts to encourage the United States to stop the effort to build the bomb. They were as dedicated to the cause of stopping the effort as they had been in starting it in 1942. But it was simply too late to turn back.

The German Effort to Build the Bomb

Werner Heisenberg of Germany was considered to be the world's greatest living scientist, second only to Albert Einstein. He was not Jewish, but he had worked with Einstein and other Jewish scientists in the past, so the Nazi machine was checking him out, threatening his future.

Heisenberg's mother was a friend of Heinrich Himmler's mother, and Mrs. Heisenberg appealed to Mrs. Himmler to have her son intervene.

At the request of his mother, Heinrich Himmler, who was head of the Nazi SS and chief of the German police, had a complete investigation performed which found that Heisenberg was a true supporter of the Third Reich and had since childhood looked for the führer who would take Germany to what he considered its rightful place in world leadership.

Himmler wrote Heisenberg with the "good" news but cautioned him to avoid such close collaboration with Jewish scientists in the future. Heisenberg could use the results of Einstein's research, but he was warned that he should always disavow Einstein as a person.

In 1939, soon after the start of the war, Germany initiated a War Weapons Bureau to develop and supply German troops with war weapons; Heisenberg was one of the first to offer his services. Within a few months, in early 1940, Heisenberg had made a proposal to develop a nuclear reactor which could lead to destructive new weapons.

He was very familiar with the work of Lise Meitner and Otto Hahn, and had duplicated their work in splitting the atom at his university in Leipzig. Heisenberg fully understood Einstein's theories, particularly as they related to the energy created when an atom was split. When a million or so atoms are split, a considerable amount of energy is released in the form of heat, causing considerable commotion.

In the beginning, Heisenberg was able to get an adequate amount of uranium from the mines in Czechoslovakia, where the richest deposits in Europe were located, and hoped to get other supplies of the essential ingredient from the Belgian Congo.

Heisenberg's first attempts for a nuclear reaction failed as he learned it took more than a pile of uranium to start a sustainable nuclear reaction. In researching the work of Enrico Fermi, he found that most of the neutrons being released were too fast to cause another atom to split, which is the basis of a chain reaction.

As an atom is split, it releases neutrons in addition to energy, and these released neutrons proceed to split other atoms ad infinitum. The released neutrons may be fast or slow neutrons. A fast neutron may speed right by a target atom and escape into space, while a slower neutron, one that is somewhat slower than the speed of light, might appear a little wobbly and more likely collide with an atom suitable for splitting.

It appeared there were many more fast neutrons than slow neutrons, and some of these fast neutrons needed to be slowed to sustain the nuclear reaction. It was determined that a slow neutron colliding with a uranium-235 nucleus will produce fission but will not in any other naturally occurring nuclide. Fission is possible, by slow neutrons, in artificial nuclides such as plutonium-239 and uranium-233. A search to determine the mechanism to slow the fast neutrons became a critical problem to be solved.

The search for a suitable moderator led to the study of the work of Harold Urey who had determined that all water in all the oceans and lakes is not simply H_2O. Occasionally there is a molecule which is heavier because it has a neutron in the nucleus in addition to the proton. This is called deuterium or heavy hydrogen. Deuterium is about twice as heavy as hydrogen. Urey speculated that there is approximately one molecule of deuterium for each ten thousand molecules of hydrogen. The process of separating these heavy molecules from water is a time-consuming and expensive task. When separated the product is called "heavy water," and is an excellent moderator for use in a nuclear reactor. (Graphite could be used equally well as a moderator. Some believe that Heisenberg looked at graphite but made a mistake in calculating its properties and ignored its use as a moderator, relying completely on the availability of heavy water.)

The Germans had plenty of raw uranium and plenty of slave laborers to do the difficult and dangerous job of machining the raw uranium to a usable form for a nuclear reactor. The task of getting enough heavy water was more difficult.

A hydroelectric facility located in a mountainous region ninety miles from Oslo, Norway, was the world's largest source of heavy water for laboratory use at that time. Obtaining heavy water from ordinary water requires enormous amounts of electricity, and the hydroelectric plant at Vermork, which utilized the 470-foot waterfall located there, provided the necessary electrical energy.

Norway came under the control of Germany, and when the German army took over the Vermork facility, it insisted on a much-increased production. Things seemed to be in place for Germany to create the ultimate weapon. With the bomb Heisenberg would create and the delivery system being developed by von Braun at Peenemünde, there was a good chance that the Third Reich could begin a thousand-year rule as Hitler had pledged.

In spite of the circumstances falling in place for Heisenberg, there were roadblocks. Hitler never really came on board as a supporter of this proposed new weapon, and Albert Speer, who was in charge of allotting all German resources, felt the war could be won with existing weapons. He believed the V-2s under development would be their ace in the hole. Speer believed that the war would be over before the scientists would be successful in completing the task of building a nuclear device. As a result of this thinking, he did not give much priority to the nuclear project. The German military was doing so well that leaders felt they needed no such weapon to conquer Europe; they could do it the old-fashioned way using aircraft and tanks, followed with the best infantry in the world. If a secret weapon were needed, the V-2 would turn the tide.

There was another factor which hindered Heisenberg's success. Almost all of the top nuclear scientists in Europe, as well as those in America, had become good friends during their visits to Göttingen in the 1920s and early 1930s. When the atom was split, many of them recognized the potential for a weapon of untold power and began to refer to this as the uranium problem. They were certainly not willing to work to give Hitler this powerful weapon, and they would do all in their power to block the German effort.

With the war well underway in late 1941, Heisenberg felt the need to visit with Bohr and discuss the possibility of the uranium weapon. Bohr knew that Heisenberg was leading an effort in Germany to explore the nuclear device. This conversation between good friends was strained and stilted due to the fear of being under surveillance.

British intelligence groups, which were so very essential to victory in the war, knew the importance of the work being done by Heisenberg and began blocking his efforts. They probably did not know of the decision of Hitler and Speer relative to their priorities. The British set about to slow down the heavy water facility in Vermork or put it out of commission.

The first raid by a group of commandos resulted in failure, and all were killed. A second raid by a smaller group was more successful and slowed production at the facility. The British studied the method of transport from Norway to Germany, and learned that the heavy water was loaded onto freight cars which went onto a ferry to cross Lake Tinnsjo on its way out of Norway. The lake was very deep, and the fact that the heavy water would sink to the bottom made a quick recovery almost impossible for German divers.

In 1944, the British learned the date of a ferry shipment and were able slip the date to the weekend, working with the leaders of labor unions in Norway who were part of the resistance.

The British wanted this slippage since the workers at the plant used the ferry on weekdays. There would be mainly holiday traffic on Sunday, which posed a dilemma for the British: Could civilians, perhaps many children, be sacrificed in the effort to stop Heisenberg and the German bomb?

The risk of allowing the Germans to get the ultimate weapon was too great. The Norwegian underground placed a bomb on the ferry and sunk it, taking the heavy water to the bottom of the deep lake. Unfortunately, several unsuspecting civilian passengers were killed, although the British and the Norwegian underground had been able to steer many passengers off the ferry without giving away the plan. The ferry was sunk on a Sunday. (Only one barrel of heavy water was ever recovered, part of a research effort approximately fifty-five years later.)

Some have speculated that Heisenberg's lack of engineering slowed his efforts and perhaps led to his failures. He was an excellent nuclear scientist but lacked the engineering skills to shape and construct a nuclear pile to start the proper chain reaction. Many of the European scientists such as Eugene Wigner, a Hungarian who fled to America, had basic engineering skills along with their science skills, and became essential in America's war effort.

Meanwhile, the Allied military forces moved across Europe. They knew where Heisenberg's operations were located and they set about to destroy them from the air.

Heisenberg moved his staff and equipment from Leipzig and Berlin to the little town of Hechingen in the Swabian Alps where he had moved his residence. The group acquired a brewery and textile mill for their offices and laboratories. A suitable site to build the nuclear reactor was found about ten miles away in the picturesque town of Haigerloch, famous for its lilacs and its good beer.

Underneath the castle at Haigerloch, a room had been hollowed out of the rock and used as a cool storeroom. The innkeeper leased the storeroom to the nuclear group, which immediately began the process of moving their reactor work to this man-made cave; all that could be seen from the sky was the castle.

Even though Heisenberg continued his work, he knew he would never finish. The Allies moved closer and eventually captured Heisenberg, who gracefully accepted defeat. He remained a hero in Germany when he was later released.

As one reflects on the German effort to build the atom bomb, it is easy to see that if Hitler had made a couple of decisions contrary to his dictatorial and obscene manner, things could have ended much differently. If he had made efforts to retain Einstein, Wigner, Szilárd, Fermi, von Neumann, and others; had allotted them the resources; and forced them to do his will, he may well have been in possession of the atomic bomb before the United States. Germany could have had a two-year head start; coupled with a successful guided-missile program, world domination would have been in Hitler's grasp.

There was another side to the story. Many of the nuclear scientists who remained in Germany and the occupied countries had formed a loose pact which pledged a less-than-full-scale research effort in pursuing a nuclear device. Robert Jungk, in his book entitled *Brighter Than a Thousand Suns,* summed up this feeling among the scientists:

> It seems paradoxical that the German nuclear physicists, living under a saber-rattling dictatorship, obeyed the voice of conscience and attempted to prevent the construction of atom bombs, while their professional colleagues in the

> democracies, who had no coercion to fear, with very few exceptions concentrated their whole energies on production of the new weapon.

It might be concluded that freedom is a strong motivator for the conscience.

The site under the mountain at beautiful Haigerloch remains as a museum and a reminder of the German effort to build a nuclear weapon. In the 1980s, I visited there and marveled at the simplicity of the German effort as compared with the three facilities in the United States. However, I realized many years after the fact that building an atom bomb is not all that difficult, given a few pounds of plutonium or bomb-grade uranium-235.

The Manhattan Project

In June 1942, the United States began a formal and urgent project to develop a nuclear weapon; the effort was named the Manhattan Project. The project was assigned to the Army Corps of Engineers, and Gen. Leslie Groves became the director of the top secret program.

Groves performed a magnificent job leading the effort which was such a great gamble and about which he had grave doubts that the task could be done in the short time allowed. He and his colleagues discussed and joked about how long the congressional hearings would take if their project spent all these billions and produced nothing. They joked about plans to rent a house near the congressional offices, just in case they were investigated.

From the beginning, it appeared Groves and his engineers would have a better chance of constructing the facilities than finding the nuclear scientists who understood the atom and its potential as a weapon. There were a limited number of Americans qualified for the effort, and the project would have to depend on many scientists who had escaped from war-torn Europe. There was also the big question of where the raw materials for a bomb could be procured.

England was a great help in bringing together the raw materials and several of the needed scientists. Prime Minister Winston Churchill and President Franklin Delano Roosevelt reached an agreement at a meeting in Quebec that the bomb project would be a joint effort; England would have an equal seat at the table in deciding when and how to use any resulting weapon. For some time, England had an ongoing project, code name "Tube Alloys," to study methods of separation of uranium-235 (U-235), including the method of gaseous diffusion. The eventual goal was to determine if a nuclear weapon could be a possibility.

In late 1943 some twenty to thirty nuclear scientists left England on a converted luxury liner and headed to the desert of New Mexico. The group from England would eventually include Niels Bohr and his son, Aage; James Chadwick; Otto Frisch; and Klaus Fuchs. (Fuchs was later found guilty of passing nuclear secrets to the USSR over a period of several years out of a weird sense of patriotic duty.)

The first person discussed for the job of leading the scientific group to develop the bomb was Ernest Lawrence of the University of California at Berkeley, who was considered to be a good man at getting things built and was deeply involved with nuclear research at the university. Lawrence was a blunt man and considered by some to be a poor risk at leading a group of brilliant and sensitive scientists. Besides, Lawrence was also considered to be essential to the difficult task of enriching uranium to create bomb-grade material.

Groves, who was somewhat like the bombastic Lawrence, then made an appointment completely out of character for him. He appointed the highly sensitive genius Robert Oppenheimer to be in charge of the scientists at Los Alamos, New Mexico, and the overall technical leader of the Manhattan Project. Oppenheimer was working in nuclear research at the University of California and was warmly endorsed by Lawrence.

The selection of Oppenheimer, who was relatively young, not in good health, and an introvert with peculiar personality traits, appeared to be a very poor choice. However, it is doubtful that there could have been a better alternative, or a more unusual pair than Groves and Oppenheimer. Early on, they generated a genuine respect for each other, and this odd couple turned out to be a team that was perfect for the gigantic job which faced them.

J. Robert Oppenheimer was born in New York in 1904, into a well-to-do Jewish family. His father was Julius Oppenheimer, a wealthy textile importer, and his mother was Ella Friedman, an artist. The family lived in a sumptuous Manhattan apartment whose walls were decorated with paintings of famous European artists such as van Gogh and Gauguin.

Oppenheimer went to Harvard in 1922, graduated summa cum laude in 1925, and then studied under Max Born at the University of Göttingen, where he received a PhD in 1927.

During the 1920s, Göttingen was the watering hole for the most brilliant physicists in the world, and great discussions and controversies about the makeup of the atom took place almost daily. During this period, a fraternity and a friendship of the most knowledgeable nuclear scientists emerged, continued throughout the war, and transcended national boundaries of the war-torn world.

After completing his PhD at Göttingen, Oppenheimer returned to the United States and became a faculty member at both the California Institute of Technology and the University of California at Berkeley. He taught classes at both universities and shared his research time between the two institutions.

In late 1942, the first nuclear chain reaction was successfully implemented by a team led by Enrico Fermi, a native of Italy who had in 1938 been awarded the Nobel Prize for his work with uranium and the subsequent fission process. Fermi and his family went to Sweden to accept the award and used that opportunity to escape Europe and move to America, where he became an honored professor at the University of Chicago.

The path to the nuclear bomb, which began in the years 1898 to 1905 with the curiosity of Becquerel and Curie, followed by the studies and conjectures of Rutherford and Bohr, the discovery of the neutron by Chadwick, the splitting of the atom by Hahn and Meitner, and culminating with the chain reaction of Fermi, was finally at hand. The chain reaction helped to convince some in the Roosevelt administration that a nuclear weapon might be possible.

In 1942, Fermi decided to proceed with his long-held plans to build an atomic pile and bring about the first sustained nuclear chain reaction. He chose the squash court located underneath the abandoned football stadium at the university as the site for his experiment.

A nuclear chain reaction is caused when a heavy atomic nucleus, like uranium, is split by a neutron causing fission, releasing energy and other neutrons which proceed to be captured by other heavy nuclei. The goal of a successful chain reaction is to build the pile (from fuel, moderator, and control) so that it becomes critical. A pile is critical when there are as many, or more, neutrons being created as are captured or lost. (For many years, a nuclear reactor was simply called a nuclear pile, since that is what it really was—a pile, or stack, of graphite blocks with lumps of uranium and control rods scattered throughout.) The process is to stack the fuel and moderator, while continually measuring the neutron creation, until the pile becomes critical. A great deal of calculation takes place before the pile is built, and in 1942, all the calculations were done without the benefit of computers.

Construction of the components of the Fermi pile began in November 1942, with the preparation of the three necessary materials. The graphite blocks to be used as the moderator had to be machined into the right configuration; the uranium oxide pellets serving as the fuel had to be pressed into the right shape; the proper instruments for measurement and control (Geiger counters and cadmium rods) had to be constructed; and a lead shield was built around the area.

The control rods inserted at the start of the pile served to absorb the neutrons being released, thus preventing a chain reaction. Day by day the pile grew, and the tension among the participants became more intense.

On the early morning of December 2, 1942, Fermi decided the pile was right for a try to go critical. The test began at 9:45 with a gradual pullout of the control rods and a measurement of the neutrons released and absorbed. The procedure was halted for the group to take time out for lunch, then the test resumed.

At 3:20 in the afternoon, Fermi was satisfied that the pile was ready to go for a sustained chain reaction and the control rods were removed. The counters which measured the freed neutrons and the fissions began to accelerate until there was a steady buzz and the first successful chain reaction occurred.

The control rods were placed back in the pile, and as the buzz died down, there was a great celebration with a bottle of Italian Chianti which had been saved by Wigner to toast the great event. Several other prominent scientists involved in the first chain reaction included Arthur Compton and Leó Szilárd.

Groves, his staff, and the scientists decided there should be three major facilities created to satisfactorily complete the nuclear bomb project. All three sites should be in relatively obscure areas where thousands of acres of undeveloped land could be procured. Two of the sites were to prepare fuel for the bomb, and the third site would be where the scientists would design and put the bomb together.

A relatively dry and warm climate where outside work could take place on a year-round basis was needed for the bomb development site. The state of New Mexico was chosen as the proper area, and several locations there were proposed and evaluated. A site where Oppenheimer had spent some time at a boy's camp was well suited for the project and was finally chosen. The site near Los Alamos, with Santa Fe as the nearest large city, would become world famous, and every nuclear scientist in the free world would work there or become some way involved.

During his first visit to the remote site, Groves speculated that he would have to build facilities for perhaps one hundred scientists and their families as well as facilities for another one hundred engineers and their families. His first estimate was off by a large factor; there would be six thousand people living and working in Los Alamos within a year.

The job of recruiting the needed scientists to Los Alamos fell to Oppenheimer, who used all the techniques in his arsenal to attract the right people. He used the arguments of patriotism, fear of Adolf Hitler, the wonderful mountain air of Los Alamos, and the thrill of working with the most brilliant group of scientists ever assembled—and he eventually recruited everyone he wanted.

Los Alamos became the home of the largest group of nuclear scientists ever assembled, and there were visits by every famous nuclear scientist in the free world. Visitors included Niels Bohr from Norway (considered by some to be the father of nuclear physics), Harold Urey, Eugene Wigner, Enrico Fermi, James Conant, John von Neumann, Leó Szilárd, and more than one hundred others.

From the nuclear scientists he had recruited, Groves quickly learned of the enormous resources necessary to succeed. They required large amounts of uranium, graphite, and heavy water, along with lots of physical space. The facilities required to get bomb-grade material from the raw uranium boggled the mind. There was also the problem of building an operating nuclear pile to produce the plutonium, another type of fuel needed for a bomb.

Through various channels and with the full cooperation of England and a chance meeting with Edgar Sengier, managing director of a major mining company named Union Minière, the United States and Britain were able to get access to most of the uranium in the Congo. Sengier had been impressed by Sir Henry Tizard, a noted British scientist, who told him of the German threat and the importance of the uranium ore his company controlled. Sengier later made arrangements to have all the available ore from their company shipped to the United States.

Oak Ridge, Tennessee, and Hanford, Washington, were the sites chosen to provide fuel for the bomb. The area chosen in East Tennessee named Oak Ridge was bounded on the south by the Clinch River and on the north by Black Oak Ridge. The area, consisting of fifty-nine thousand acres, was located twenty-five miles west of Knoxville, home of the expanding Tennessee Valley Authority.

Some early visitors considered the area and the people quaint, perhaps backward, and left behind as technology advanced; they soon learned that the residents were hard working, honest, and patriotic.

The area which lay in the counties of Anderson and Roane consisted of the four crossroad communities of Scarborough, Wheat, Robertsville, and Elza. The citizens (approximately one thousand in number) were evacuated for security reasons, some giving up their ancestral homes. They were given court orders to vacate on a very short notice and many felt they were not properly compensated for their property or given sufficient notice. It was a period of extreme urgency and the government operated under the guideline that the U.S. was in a fight for survival.

The area chosen in the state of Washington was along the Columbia River, which provided large quantities of water and electricity. The area was composed of five hundred thousand acres in a remote area which provided good cover for security. Furthermore, the very remote area was necessary for building the nuclear reactor in the event of a nuclear accident. The three small Washington towns of Hanford, White Bluffs, and Richland were located in the chosen area, and like the towns at the Oak Ridge facility, they vanished as fifteen hundred people were moved to other sites. The new facility was named Hanford Engineer Works.

The site at Hanford was used to build the nuclear reactor to generate fuel for the plutonium bomb. Generating plutonium as fuel for the bomb was favored by the European scientists and was the method Heisenberg was using in an attempt to build the German bomb. The method has been compared to the hopes of alchemists of old who thought they could convert lead into gold. An operating nuclear pile would create plutonium as a by-product of the nuclear reaction taking place in the pile; however, the separation of the plutonium from reactor remains was a problem not then solved. It was believed by most of the European scientists that the plutonium bomb would be easier and more assured of exploding as planned as compared with the uranium bomb.

The American scientists led by Ernest Lawrence preferred the method of enriching uranium—separating U-235 from raw uranium and using the U-235 as fuel. The site at Oak Ridge was given this task, and they used the two methods of separation which the nuclear scientists considered most likely to succeed.

The process of enriching raw uranium to obtain more U-235 was thought to be possible in various ways which had never been attempted in quantity. The two most likely methods of separation were thought to be through gaseous diffusion and electromagnetic means. The method of gaseous diffusion requires that the raw uranium be in a gaseous state, known as uranium hexafluoride, and the resulting separation process relies on the difference in weight of the isotopes. Uranium-235 has three less neutrons than uranium-238. In the gaseous state and under pressure, U-235 is in a more rapid random motion than the heavier U-238. A good description of the process is given in the book by Gen. Leslie Groves entitled *Now It Can Be Told: The Story of the Manhattan Project*:

> The method was completely novel. It was based on the theory that if uranium gas was pumped against a porous barrier, the lighter molecules of the gas, containing the U-235, would pass through more rapidly than the heavier U-238 molecules. The heart of the process was, therefore, the barrier, a porous thin metal sheet or membrane with millions of submicroscopic openings per square inch. These sheets were formed into tubes which were enclosed in an airtight vessel, the diffuser. As the gas uranium hexafluoride was pumped through the long series, or cascade, of these tubes, it tended to separate the enriched gas moving up the cascade while the depleted moved down. However, there is so little difference in mass between the hexafluoride of U-238 and U-235 that it was impossible to gain much separation in a single diffusion step. That was why there had to be several thousand successive stages . . .

The diffusion process may require more than a thousand stages to reach the proper concentration of U-235; extensive heating and cooling processes were needed during each stage, and in fact, throughout the entire facility. This process had never been done in any quantity and would require the most sophisticated engineering and construction yet imagined, as well as extensive energy resources. The Oak Ridge building at the gaseous diffusion plant at the K-25 facility was more than a mile long, and at the time of construction, was the largest building in the world.

Considering the infinitesimal dimensions of the isotopes involved, it is unbelievable that anyone could even consider the separation process proposed for Oak Ridge. Lawrence, however, had confidence it would be successful.

The electromagnetic process of separation is based on the principle that an ion travels a curved path as it passes through a magnetic field. If the magnetic field is of constant strength, the heavier ions will describe curves of longer radii. Therefore, the theory is that isotopes of different mass, such as U-235 and U-238,

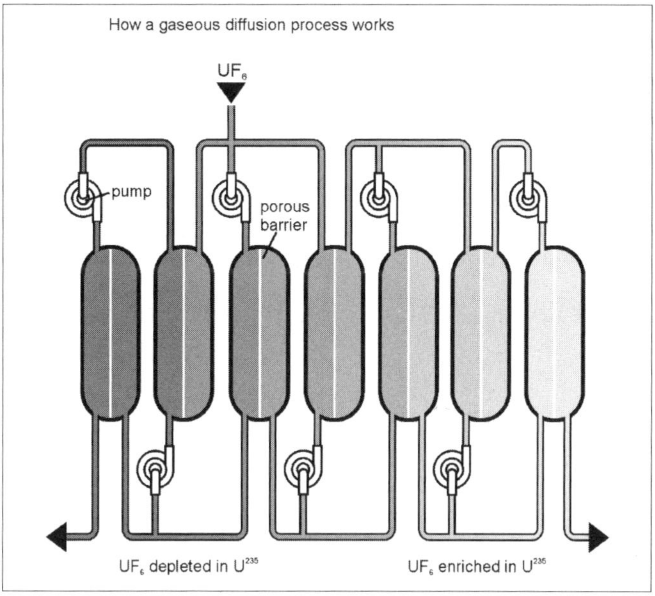

Author's illustration of a gaseous diffusion schematic.

can be isolated and collected. A process, requiring massive magnets and metal resources, had to be constructed to collect the U-235 isotopes from their shorter radii. Groves describes the process in *Now It Can Be Told*:

> Electromagnetic isotope separation enlarged and elaborated Francis Aston's 1918 Cavendish invention, the mass spectrograph. As a 1945 report by the Lawrence staff explains, the method "depends on the fact that an electrically charged atom traveling through a magnetic field moves in a circle whose radius is determined by its mass"—which was also a basic principle of Lawrence's cyclotron. The lighter the atom the tighter circle it made. Form ions of a vaporous uranium and start them moving on one side of a vacuum tank permeated by a strong magnetic field and the moving ions as they curved around would separate into two beams. Lighter U-235 atoms would follow a narrower arc than heavier U-238 atoms; across a four-foot semicircle the separation might be about three-tenths of an inch. Set a collecting pocket at the point where the

> U-235 ion beam separately arrived and you can catch the ions. "When the ions strike the bottom of the collecting pocket . . . they give up their charge and are deposited as flakes of metal."

Both of these methods proved to be very difficult and time consuming, and required major resources. Someone once said that if these two ways were the most likely to succeed, the others had to be impossible. One scientist in charge of the gaseous diffusion process described the problem as difficult as trying to find a needle in a haystack while wearing boxing gloves! The tasks assigned to both Oak Ridge and Hanford required enormous amounts of electric and water resources, tons of materials not yet manufactured, and numerous trained specialists.

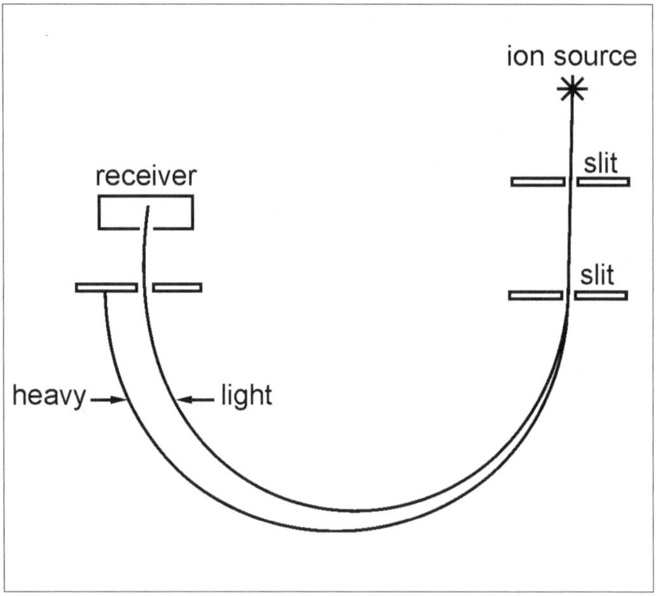

Author's illustration of the magnetic field process.

In a very few months, thousands of engineers, scientists, and laborers were working at Oak Ridge and Hanford. Their task was to produce as quickly as possible U-235 and plutonium in a quantity sufficient to build a nuclear bomb. In the meantime, scientists at Los Alamos explored means to build a bomb once the fuel was ready.

There were different approaches for using U-235 and plutonium, due to the big difference in the detonation system required. Plutonium is much hotter and easier to start a chain reaction than U-235; this requires different methods of detonation be used for the two bombs.

By late summer of 1945, both facilities produced enough fuel for a bomb. The group at Oak Ridge produced slightly more than the sixty-four kilograms of enriched U-235 that was used in the Hiroshima bomb. There was not enough U-235 to use for tests, so the U-235 bomb would have to go cold turkey.

After two years of considering it, the scientists had more confidence in the detonation process of the U-235 bomb than in the plutonium bomb. The procedure of implosion to be used with the plutonium bomb was thought to be more of a gamble.

Since there was an adequate supply of plutonium, a test was planned and carried out with good success at Alamogordo, New Mexico. The test at Alamogordo was a significant event in building the bomb and took place with the most intense planning and precaution of almost any experiment in history.

For several reasons, the main concern was about the weather. The wind direction was important because of radioactive fallout, particularly over populated areas, and it was best if it did not rain since this would bring the radioactive fallout down with it. Also, the weather had to be good for flying since aircraft were to be in the air to take readings and photographs.

Groves and Oppenheimer took many divergent opinions, mixed them together, and then took some chances, so the test was allowed to proceed. The first atomic blast took place at 5:30 AM on July 16, 1945. This initial test convinced Groves and others in the government that the bomb would work and could lead to an early end to the war.

With the fuel in their possession, scientists at Los Alamos had to construct the bomb and determine a method for detonation.

The method devised for the U-235 bomb was to divide the available concentrated uranium into two units and propel one unit into the other with a force large enough to start the nuclear reaction. The larger unit weighed thirty-eight kilograms and was the bomb; the remaining unit comprised the bullet. Ordinary explosives were used to propel the bullet down a gun barrel with velocity adequate to start the nuclear reaction.

The construction of the gun barrel and its weight and thickness was a matter of concern to these nuclear geniuses until a practical engineer suggested that the gun barrel could be much lighter than that of ordinary cannons because it would be used only once, there would be no recoil, and the whole thing would be vaporized by the nuclear blast.

The bullet had to impact the bomb with sufficient velocity in a manner that would ensure a complete nuclear explosion. The geometry used in preparing the bomb was very important. One might imagine the bomb part of the U-235 formed into a solid like a baseball catcher's mitt, and the bullet would hit it right in the pocket (which was configured to bring about a complete detonation) with velocity to start the reaction and resulting nuclear explosion. Much thought was done relative to the shape of the catcher's mitt so that most, if not all, of the full amount of the U-235 would be part of the reaction and not be destroyed by the original blast. The uranium bomb was long and slender due to the nature of the trigger device and was named the Little Boy bomb.

Due to the more explosive nature of plutonium (if two pieces of plutonium are fired at each other, the plutonium starts exploding before the two parts collide) a different technique is necessary. A method was devised by Oppenheimer and his team of scientists to start with a ball of plutonium of low density and then wrap high explosives around this ball to cause an inward explosion. The resulting cascade gives the ball of plutonium time to set the proper reaction before it explodes. This method is called the method of implosion, and the procedure proved workable with the test at

1. Box tail fins
2. Steel gun breech assembly
3. Detonator
4. Cordite (conventional) explosives
5. Uranium-235 “projectile,” six rings (26kg) in a thin can of steel
6. Baro sensing ports and manifold
7. Bomb casing wall
8. Arming and fusing equipment
9. Gun barrel, steel, around 10 cm diameter, 200 cm length
10. Arming wires
11. Tamper assembly, steel
12. Uranium-235 “target,” two rings (38 kg)
13. Tamper/reflector assembly, tungsten carbide
14. Neutron initiator
15. *Archive* fuzing radar antennas
16. Recess for the boron safety plug (not shown) to be ejected into

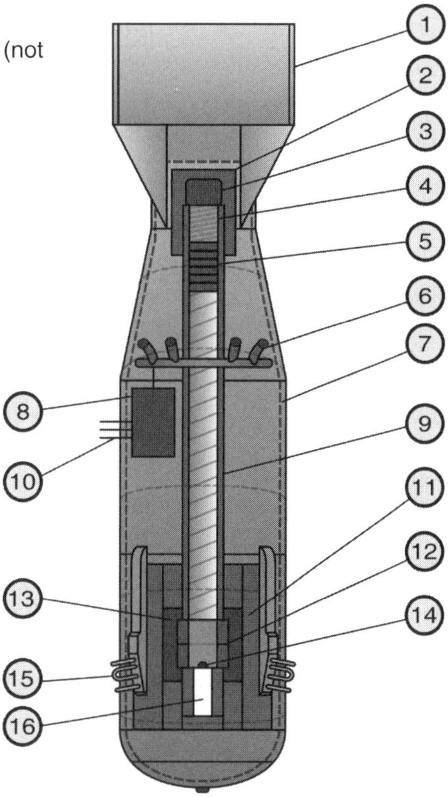

Diagram of the Little Boy bomb. Courtesy of Wikimedia Commons.

Alamogordo. The plutonium bomb was like a huge beach ball or perhaps a large egg, and was given the name Fat Man.

Little Boy and Fat Man exploded as planned at Hiroshima and Nagasaki, and ushered in the nuclear age.

There have been many heated discussions and debates concerning the decision to build the atom bomb. Many of the scientists who were involved in building the bomb have expressed regrets at having been part of the effort. Szilárd, Wigner, and others who had worked so hard to get the United States to build the bomb changed their minds when Hitler was defeated, and tried just as hard to discourage the United States from completing the project. It was too late.

Since the end of World War II, threat of the use of nuclear weapons has been paramount in the minds of most world leaders. The United States and the USSR went through more than twenty-five years living under the umbrella of Mutual Atomic Destruction (MAD). Non-Proliferation (atomic weapons) treaties have been signed by almost all the world's nations. However, there is always a concern that some rogue nation will develop nuclear weapons and defy the rest of the world. Leaders still search for ways to control the nuclear weapon.

THE NUCLEAR AGE: FEAR AND HOPE

In late 1944, Gen. Leslie Groves and his staff initiated the plan to drop the atomic bomb on the mainland of Japan. During the planning process, it was necessary to set up positions in the Pacific within fifteen hundred miles or so of Japan, requiring extreme coordination with United States commanders, while at the same time keeping the atom bomb a secret.

The area around Saipan and Tinian was chosen as the site from which the bombing runs would be launched, and the airfield at Tinian was considered best suited to host the Boeing B-29 bombers to be used for the raid. There were many test runs by the B-29s with the dropping of a peculiar-shaped dummy of the Fat Man or the Little Boy, and the crews of the test runs who had no idea of this new-type bomb thought the folks back in the United States were going crazy.

The Little Boy bomb (U-235) was transported from San Francisco to Tinian aboard the heavy cruiser USS *Indianapolis,* and arrived at Tinian on July 26, 1945. After unloading its precious cargo, the *Indianapolis* proceeded to Guam, then departed shortly thereafter for the Philippine Islands. The *Indianapolis* was to spend several weeks in the Philippines, training its crew of 1,196 men so they could join Task Force 95 at Okinawa for the invasion of Japan, scheduled for November 1, 1945, at Kyushu.

The *Indianapolis* never made it to the Philippines. On July 30, while crossing the Philippine Sea, the cruiser was struck by three torpedoes from the Japanese submarine *I-58.* The *Indianapolis* was hit in critical areas, causing explosions on board from stored ammunition, as well as explosions from boilers in the engine room. Initial explosions knocked out the power center and left the ship helpless to communicate. The ship took a 45 degree list, and the spilling fuel started fires.

It was estimated that approximately 850 men were able to abandon ship or were thrown into the water, while the remaining 350 men went down with the *Indianapolis.* More than 50 died during the first night in the water, and with the dawn, came sharks. One survivor counted more than 25 attacks by sharks on the first day in the water. And another, the ship's doctor, Lt. Cdr. Lewis Haynes, said he counted 85 shark attacks during the nearly four days the crew of the *Indianapolis* fought to survive. During the eighty-four hours in the water, more than 500 men had died. There were only 318 survivors of the original crew of nearly 1,200.

The *Indianapolis* had been a participant in the pivotal battle at Okinawa in April 1945. The U.S. Navy sustained bigger losses at

Okinawa than any other battle in its entire history, losing five thousand sailors, thirty-six ships, and nearly eight hundred aircraft. The total loss of U.S. sailors, soldiers, and marines was twelve thousand. The Japanese lost some one hundred thousand, and estimates were that one hundred thousand civilians were also killed. Okinawa was the bloodiest battle in the Pacific, and was, of course, fought on Japanese homeland.

The fierce battles at Okinawa following the heavy losses on the beaches at Iwo Jima were fully evaluated by the Washington leaders when they considered whether to proceed with their plan to drop the first atomic bombs. There were intense disagreements in the group surrounding President Truman, who had this great problem dumped in his lap upon the death of President Roosevelt. Some of the military leaders believed Japan was finished and would soon surrender, while others predicted the Japanese would fight on the beaches with pitchforks and fists to the death. They knew that all schoolchildren in Japan spent two hours a day practicing with sharpened bamboo sticks to attack and repel the invaders. James Byrnes, who was later named secretary of state, led the group arguing for dropping the bomb which, in the end, carried the day.

Hiroshima was selected as the primary target for Little Boy, the first nuclear bomb. Kokura, home of a large arsenal, was the secondary target, and Nagasaki was the second alternate. The plan called for a visual bomb drop, but if this was not possible, the secondary target would be selected.

The B-29 named the *Enola Gay* was to carry the bomb, and would travel without fighter escort to avoid attention. It was hoped that if it were spotted it would be evaluated as an observation plane doing reconnaissance. The *Enola Gay* was captained by Col. Paul Tibbets, selected as the best man in the air corps for this mission.

The bomb was set to explode at nineteen hundred feet above the city. As he had done in so many other instances, John von Neumann made the calculations for the optimum altitude for the detonation. This was before computers were generally available (the Mark 1 at Harvard was used for some of the calculations relative to the bomb), and it was extremely fortunate that von Neumann, Enrico Fermi, and some others were there to do the complicated computations.

The weather report for the area over Hiroshima was not good. As the *Enola Gay* approached Hiroshima, however, the area over the city cleared and the visual sighting became possible. The flight of some seventeen hundred miles was made right on schedule by Tibbets and the crew of the *Enola Gay*.

The bomb was dropped on Hiroshima at 9:15 AM (local time) on August 6, 1945, within a few seconds of the scheduled time for the drop. The crew recorded several visual observations of the mushroom cloud's height and width, and the *Enola Gay* was hit by two shock waves, the first being a jolt big enough to cause real concern. There was some jubilation by the crew, tempered by the awesome sight they had just witnessed. The return to Tinian was made without incident.

The city of Hiroshima was destroyed. The Little Boy bomb, with a total weight of four tons and containing approximately sixty-four kilograms of U-235, had the explosive power of 12,500 to 20,000 tons of TNT. The bomb detonated at 1,900 feet above the city just as planned, and the location of the impact was 550 feet from the aiming site. At the time, the population of the city was estimated to be 290,000 civilians plus some 45,000 soldiers. Later estimates were that more than 100,000 were killed instantly, with another 50,000 dying within a few months. All the infrastructure of the city was demolished.

There was some restrained celebration and congratulations in the military and among the war planners. The end of the war was at hand, but most of the scientists were saddened by the massive devastation of their creation and wondered how this monster could be controlled in the coming years.

The bombing of Hiroshima caused intense pressure for Japan to surrender. The civilians wanted to sue for peace, but the military was not ready to accept unconditional surrender.

American leaders decided to go ahead with the Fat Man bomb which was already in Tinian being readied for the drop. The primary target for the Fat Man was to be Kokura Arsenal on the island of Kyushu, with the secondary target being the port city of Nagasaki. The B-29 which would carry the Fat Man was named the *Bock's Car,* and was piloted on this occasion by Maj. Charles Sweeney.

The *Bock's Car* left Tinian shortly before dawn on August 9. When the plane reached Kokura there were clouds and smoke over the city and Japanese fighter planes were preparing to check out the B-29, so the decision was made to go directly to Nagasaki and drop the bomb.

Upon arrival at Nagasaki, the plane found thick clouds, but a hole opened up at the last minute and the bomb was dropped at 11:02 AM, exploding 1,650 feet above the city. The explosion was contained by the surrounding hills, so the devastation was not as bad as at Hiroshima. The city was destroyed, however, and some 74,000 had died at Nagasaki by the end of 1945.

The *Bock's Car* had barely enough fuel to return to Okinawa, and Sweeney reported he did not have enough to taxi off the runway.

Even the second bomb was not enough to persuade Japanese military leaders who wanted to fight to the death rather than surrender. The U.S. was preparing to ship a second plutonium bomb while awaiting word from Japan. The emperor finally intervened and accepted the terms of the Allies—with the condition that Japan would retain its emperor, a term finally accepted by the United States. The emperor announced on August 15 that he had accepted the terms of surrender, and the nation of Japan wept while accepting the decision.

The Japanese navy was almost completely destroyed at the end of the war, and the Japanese armies had been defeated and decimated in the Philippines, as well as at Iwo Jima and Okinawa. Japan was a defeated nation with no chance of victory. The Japanese military leaders should have followed the desire of the civilian leaders and accepted the surrender offers before the bombs were dropped on Hiroshima and Nagasaki.

Chapter 7

Oak Ridge and the Oracle

When we moved to Oak Ridge in 1954, the town still resembled an army base. The houses of wood-frame construction gave the appearance of being quickly erected, and there were large dormitories for singles and couples without children. The small shopping centers were somewhat like the post exchanges on a military post, and a large street called the Oak Ridge Turnpike ran from east to west, splitting the town in half. Highway 25W from Knoxville to Clinton bounded the area on the east and Highway 62 from Knoxville to Oliver Springs ran through the middle of the town from south to north.

There were still three major research and production facilities in Oak Ridge. The facility called K-25 was a very large plant used for enriching uranium (obtaining U-235) through the process of gaseous diffusion. A facility called Y-12, established to enrich uranium by electromagnetic means, had also been charged to look at the dangers and effects of radiation; this was being referred to as the new field of health physics.

The third facility, originally named X-10, was renamed the Oak Ridge National Laboratory (ORNL), and had become the research facility established to design and build nuclear reactors and to analyze chain reactions.

In early 1943, Enrico Fermi decided there should be an air-cooled graphite and uranium reactor ("reactor" gradually replaced the word "pile") built in Oak Ridge. This would be in addition to the two facilities, K-25 and Y-12, to be used for the separation of uranium-235 (U-235) from uranium-238 (U-238). The Oak Ridge facilities, originally called the Clinton Engineer Works, added an area called X-10 as the location to build the reactor.

The Dupont Chemical Company was contracted to construct the reactor at X-10, and the University of Chicago was brought in to operate the facility. It took some cajoling and arm-twisting to recruit since Oak Ridge was considered to be in the backwoods and not an easy place to get scientists and engineers to relocate. In fact, there was no place to get a legal adult beverage and the town was called "Dogpatch" just as often as it was called Oak Ridge. However, one of the best arm twisters during World War II was Gen. Leslie Groves.

Excavations for the graphite reactor began in April 1943, and during the summer, work was completed on laboratories for chemistry, physics, machine shops, warehouses, and administration buildings at the X-10 site. The foundation and walls of the reactor were constructed of very thick concrete and were completed in July. The graphite pile (graphite was used as the moderator) was encased inside this very thick concrete box and consisted of seventy-three layers of pure graphite blocks to build a pile in the form of a cube which would measure twenty-four feet on a side (13,824 cubic feet). The graphite blocks were cut and shaped on site and they were configured with grooves running through the graphite which would allow loading the uranium with long prods and the insertion and withdrawal of control rods.

Control rods, generally made of cadmium, would serve to control the pace of the nuclear reaction by absorbing neutrons before they had an opportunity to collide with a fissionable target. The rods could be withdrawn at the proper time to accelerate the reaction.

The uranium was configured in slugs and wrapped in aluminum packets provided by ALCOA. The pile was to be air cooled by large fans, the world's largest at the time, which would suck off the heat and move it through a large chimney adjacent to the building.

With the graphite blocks shaped and stacked in place, the process of inserting sixty thousand uranium slugs into the graphite pile began. The job of inserting the uranium slugs needed to be done as rapidly and carefully as possible, which made loading the fuel into the reactor difficult and nerve-wracking work to be continued around the clock until completion.

The day shift loaded ten tons of uranium slugs and the night shift continued at the same or accelerated pace. Every so often, after a ton or so of uranium was inserted, control rods were extracted to measure the level of neutron flux and to attempt estimating how much more it would take to reach a critical condition.

It was determined at midnight that one more load of slugs would bring the pile to critical. It took just sixteen hours to reach this condition; only half of the 1,248 channels had been filled with uranium slugs. This had been predicted by Eugene Wigner and Alvin Weinberg, who also accurately predicted the uranium fission rate of the pile.

Arthur Compton and Enrico Fermi, both Nobel laureates, were on standby in Oak Ridge. A car was dispatched for them, and they arrived for the loading of the final slugs. The reactor went critical as predicted and gave the scientists confidence that their theoretical calculations were close to correct. The completion of the X-10 reactor on November 4, 1943, was a milestone in the development of nuclear reactors.

The X-10 graphite pile was a big success as it became the first powerful working reactor ever built, producing plutonium essential for the operation at Hanford. Before the end of 1943, scientists were extracting plutonium from the uranium slugs which were removed from the pile. Extracting the plutonium from the processed uranium was a tricky task, and new techniques had to be learned. Some modifications allowed the reactor to operate at four thousand kilowatts as the X-10 pile continued in operation for twenty years. Today, the graphite pile is on the registry of historic sites in the United States and can be seen on tours of the Oak Ridge area.

The success of the Oak Ridge graphite reactor proved the viability of the nuclear power reactor and led to many being built over the succeeding years. The first industrial power station was built in Calder Hall in Cumberland, England, in 1956.

At the time I arrived, ORNL was headed by Clarence Larson, and the research director of the laboratory was Dr. Alvin Weinberg, a noted nuclear physicist who participated in the early nuclear work at the University of Chicago.

Dr. Weinberg, a son of Russian immigrants, was educated at the University of Chicago, where his major area of study was mathematical biophysics. He was a member of the group which worked with Eugene Wigner in the area of theoretical physics at the university during the war. They studied the geometric arrangements of a nuclear pile to learn how best to attain, sustain, and control a nuclear reaction.

Dr. Weinberg came to Oak Ridge with Wigner in 1945 as head of the physics division. Within a few years, Dr. Weinberg became technical director of the laboratory, and then director of the laboratory in 1955, where he remained for many years. He promoted many peaceful uses of nuclear energy and emphasized other areas of research, expanding the mission and research reputation of the ORNL.

In 1947, Dr. Weinberg created a computation section in the physics division and hired Dr. Alston Householder from the University of Chicago to head the section. In 1948, the computation section was converted to an independent group called the mathematics panel, and was charged with doing complicated, multifaceted calculations and looking at the possibility of acquiring its own automatic computers. Until that time, all computations were done on desk calculators and some card-programmed machines (much as was done at the Guided Missile Development Division in Huntsville).

The first automatic computer acquired by the math panel was a matrix multiplier designed to facilitate the solution to sets of linear algebraic equations. In addition, the group purchased computer time on the Mark 1 at Harvard University.

The Mark 1 was constructed of switches, relays, rotating shafts, and clutches. Even though it is considered by some to be the first digital computer built in the United States, it was not electronic and is often left out of the history of present-day computers.

Dr. Householder, who had done important research in the field of mathematical biophysics, was about to enter the field of computing and error analysis where he would become one of the most famous researchers and innovators.

Householder spent his childhood in Alabama and was educated in public schools. He attended Northwestern University in Evanston, Illinois, and received his bachelor's degree in 1925. He received his master's degree in philosophy from Cornell University in 1927.

To earn a living, Householder taught mathematics at many places, and his interests gradually switched to applications of mathematics, particularly applications related to biology. He was awarded a PhD in mathematics from the University of Chicago in 1947 for a thesis based on the calculus of variations.

As computations moved into areas where precise calculations were no longer possible, it became necessary to use approximations and iterative methods to reach solutions. The general field of calculating errors in numerical procedures and devising methods to accelerate convergence to solutions came to be known as numerical analysis. Dr. Householder published several books and many research articles in this field and became a world authority in the area of mathematical computations and error analysis.

The computer could add or subtract two twenty-three-digit numbers in three-tenths of a second, and multiply time was four seconds, while divide time was ten seconds.

In the early 1950s, there was no end to speculating how nuclear energy and nuclear technology could be utilized in the coming years. There was great hope that the useful benefits of nuclear technology would overcome the many fears existing over an energy source that could destroy mankind.

One of the big projects underway was the Aircraft Nuclear Propulsion (ANP) project, with the goal to build an airplane that could fly almost forever without refueling. The nuclear submarine was a hot item, and Admiral Hyman Rickover had already spent a year in Oak Ridge, attending classes in reactor technology. There were also programs underway to study nuclear power on ships of all kinds and the use of nuclear power in space trips to the moon and beyond. The most obvious and most successful use of nuclear technology was in the area of generating electric power using nuclear reactors.

Some of these great hopes such as ANP never got off the drawing board, but some did and were great successes. Good benefits in nuclear medicine have also taken place over the years. Medical researchers were quick to make use of radioactive isotopes. These isotopes were used to study body chemistry by serving as tracers through the human body. Radioactive scans are used for a medical diagnosis through radioactive scanning, locating, and defining tumors. Radioactive elements are also being used for treatment of certain cancers.

In 1953 the mathematics panel installed the ORACLE (Oak Ridge Automatic Computer and Logical Engine) as its first large computer. In Greek mythology, oracle was often a shrine through

John von Neumann was born in 1903 in Budapest, Hungary, and received a PhD in mathematics from the University of Budapest in 1927. Prior to receiving his PhD, he published several papers on subjects in higher algebra, set theory, and quantum mechanics. Von Neumann went to Princeton University in 1930 and was appointed a permanent member of the Institute for Advanced Studies in 1933. By 1940, he was recognized as one of the world's foremost mathematicians.

Von Neumann was one of the first mathematicians to give credibility to what was generally called applied mathematics, and for this reason he was not known as a mathematician's mathematician. However, this inclination of von Neumann made him invaluable in solving many of the problems which came about due to World War II.

During the war, he became a consultant to the U.S. Navy (Bureau of Ordnance), the U.S. Army (Ballistics Research Laboratory), and the Los Alamos Scientific Laboratory. At Los Alamos, he was given the task of calculating the precise altitude at which the Hiroshima bomb should be detonated for maximum effect. His role during the war gave him a status reached by very few mathematicians in history.

Von Neumann learned of the ENIAC development in 1944, and in September of that year made visits to the ENIAC at the Moore School of Engineering. He encouraged his colleagues at Los Alamos to try some of their computations on the ENIAC, and as a consequence of his suggestion, one of the first tests of the ENIAC was an application from Los Alamos.

Planning for the follow-on contract of the EDVAC (a stored-program computer) was given considerable direction and impetus by von Neumann. In June 1945, von Neumann published a report entitled "First Draft of a Report on the EDVAC." This report is still considered to be the fundamental statement of the plan for a stored-program electronic digital computer.

There were discussions prior to this time for a stored-program computer. As a matter of fact, the ENIAC was planned as a stored-programmed computer, but the plan was dropped due to the press of time. There is no evidence the EDVAC was ever completed.

A wonderful book by Nancy Stern entitled *From ENIAC to UNIVAC* (Digital Press, 1981) gives a complete history of these developments and contains the complete fundamental paper by von Neumann as an appendix. Stern's book also gives an excellent overview of the controversies surrounding von Neumann's paper and of the patent claims by J. Presper Eckert and John Mauchly.

In the period following the development of the ENIAC, there was considerable discussion on university campuses as to whether or not a professor should profit financially from patents which resulted from research efforts. The Moore School established a policy requiring all professors to agree that patents belonged to the school. As a result, Eckert, Mauchly, and several other distinguished professors left the school.

Von Neumann returned to the Institute for Advanced Studies (IAS) at Princeton but maintained his interest and leadership role in the development of digital computers. The development of a computing tool to calculate results for practical applications was not in the scope of effort planned for the IAS. However, the leadership at IAS decided that a few rules could be bent (or even broken) for a man such as John von Neumann. From his start with the ENIAC and EDVAC, the concepts laid down by von Neumann led to the development of many computers.

which a deity reveals hidden knowledge. The ORACLE came from a line of computers whose design was led by John von Neumann and consisted of computers with the names ILLIAC, ORDVAC, AVIDAC, and JOHNIAC.

The ORACLE was an improved and faster version of the AVIDAC (Argonne's Version of the Institute's Digital Automatic Computer), which was built for use at the Argonne National Laboratory in Chicago. Dr. Householder later contracted with Argonne to build an improved AVIDAC for Oak Ridge. Engineers Earl Burdette, Bud Klein, Jim Alexander, James Woody, and Dr. Householder were added to the team for the ORACLE development. The ORACLE became fully operational in the summer of 1953 and was moved to Oak Ridge in October.

The ORACLE was a binary machine with a word length of forty bits. (At the time, the word *byte* had not yet been invented.) In the ORACLE, all numbers were assumed to be between +1 and -1 (called a machine number), and all numbers used in computations had to be scaled so that numbers resulting from computations would be machine numbers; otherwise an overflow would occur, and the results would be meaningless. This was called fixed decimal point arithmetic.

All of the arithmetic in the ORACLE was carried out in the arithmetic unit. For two numbers to be added, one of them had to go into the A register, which was part of the arithmetic unit, then the other number was added to the current contents of the A register; the result was left in the A register. The forty-bit word was broken down into ten four-bit characters. Each one of the four-bit characters represented a hexadecimal (number base of sixteen) digit (0, 1, 2 . . . 9, A, B, C, D, E, F).

Programming was done using an ORACLE word for two instructions. The first two hex digits were the instruction, and the second three were the storage location referred to by the instruction. The instructions to add two numbers (written in hexadecimal) would have been:

CA021 AD348 AF453

These three instructions would add the contents of storage location 021 (the command CA021 cleared the A register and fetched the contents of storage location 021 into the A register) to the contents of memory location 348 and placed the result in storage location 453.

An add instruction used one-and-a-half ORACLE words. For subtraction, complement gates replaced the subtrahend by its complement and then addition proceeded as before. Multiplication was accomplished by a sequence of additions and shifts.

Division was done very much like it would be done with pencil and paper. The dividend was sent to the A register. (The ORACLE had two registers, A and Q, which could shift together. It was possible to shift A and not Q, but not possible to shift Q and not A.) The divisor was subtracted from the current remainder in A. This new trial remainder was accepted if its sign was the same as the old one. In this event, the final bigit (binary digit) was made 1, and both registers shifted left. Otherwise, the new remainder was rejected, and the final bigit in Q was made "0" before the shift took place.

So this was it. The entire arithmetic unit of the ORACLE consisted of an adder, a shifter, some storage registers, and the supporting circuitry. Many of the ORACLE's built-in operations were logical in character, rather than arithmetic. One such logical operation was the transfer of control to another location where other program instructions were stored.

The memory in the ORACLE consisted of 1,024 digits (words) using forty Williams tubes (modified cathode ray tubes). Each

Williams tube was divided into a thirty-two by thirty-two raster, giving 1,024 spots on each tube. There were forty tubes resulting in memory for the ORACLE of 1,024 forty-bit words of memory.

Actually, there were eighty tubes arranged in two banks of forty for error checking. During operation, the two banks of forty tubes received the same signals. A checking system was devised which always compared the corresponding spots. If both were 0 or 1, there was no problem. If one was a 0 and the other a 1, then the spot was called a 1 since the possibility that a spurious 1 could be generated was much less than a spurious 0.

The programmer of the ORACLE had to be careful not to write iterative loops too tight. As was the case in many instances on early computers, the programmer needed to know a considerable amount about the engineering characteristics of the hardware. The problem with iterative loops and the calculation of the read-around ratio had to do with the construction of the Williams tubes which were used as memory on the ORACLE.

I could not recall precisely how things worked, so I once again asked my friend Chuck Hubbard for help. He now lives in retirement in Pendleton, South Carolina, and replied:

> The physical principle behind the Williams tube memory rests on the fact that the coating on the inside face of the tube has a secondary emission ratio greater than one. This means that on the average, for every electron striking the coating, slightly more than one electron is released from this coating. These freed, stray electrons are then attracted to the positive plates within the tube, providing no other relatively positive target is in the vicinity. If the electron beam is focused on a single spot for a short duration of time, the spot becomes positively

charged as electrons are bled off. This condition represents a "1." If the electron beam is then focused slightly (much less than that distance to another of the 1,024 positions) away from its initial position, then some of the stray electrons produced will find their way back to the positive spot in the coating, thus changing the positive charge just created back to neutral, representing a "0."

To make a Williams tube out of a cathode ray tube, it is necessary to glue a fine copper screen to the outside face of the tube which is connected via a capacitor to a discrimination circuit which determines the state of the spot being focused upon, i.e. it is identified as a "1" or a "0." This determination is made on the shape of the pulse as the beam approaches the target spot. The pulse for the "1" is stronger, or more positive, than the pulse resulting in a "0."

In the write cycle, a single burst of the electron beam (a dot) will result in a positive charge or a "1." If this burst is followed by a repositioning of the beam (a dash) then the "1" state is returned to the "0" state.

In the read cycle, the discriminator stores the result in a binary register, but since the read cycle will always leave a "1" state on the tube, it must be immediately followed by a write cycle to restore the previous value.

The manner of construction of the early computers required that the programmer understand the engineering and placed a responsibility on the programmer that today's computers do not. Programmers for the ORACLE were required to calculate a read-around ratio whenever an iterative loop was programmed. The time to execute tight loops would have to be manually calculated to determine how many times the loop would be executed within the 20 milliseconds allowed. If the number was over the allotted 100 times, the programmer would have to do one of two things: either read the problem spot to regenerate it before executing more loops, or slow down execution time of the loop by adding superfluous instructions to the loop such as extraneous multiplications or divisions. The reason this had to be done was due to the way the Williams tube memory worked. Each spot on the tube has four neighboring spots except for the spots on the boundaries of the raster which has only three neighbors. When the neighbors of a particular spot which has a "1" state are addressed by the electron beam, some of the stray electrons produced will be attracted to this positive charge. Because of the relatively great distance between this spot and its neighbors, the number of these electrons is not sufficient to change the "1" back to a "0." However, if the neighbors are addressed often enough within a major regeneration cycle, the total of all these electrons may be sufficient to neutralize the charge or drop a "1." A major regeneration cycle consists of periodically performing a read/write on each of the 1,024 spots to assure strong positive charges for the spots in the "1" state. This major regeneration cycle occurs every 20 milliseconds.

The read-around ratio is defined as the number of times the neighbors of a spot can be addressed without addressing the spot itself, within a major regeneration cycle, without the danger of changing a "1" to a "0." The guaranteed rule of the read-around ratio of the ORACLE depended somewhat on the quality of the tube coatings, but was approximately 100. This number was periodically determined by memory diagnostic test routines.

The ORACLE had an auxiliary memory consisting of four magnetic tapes. These four tapes were each one thousand feet in length and two inches wide. The total storage capacity of the magnetic tapes was 3,600,000 words. Seek time (to find needed data) could take up to a minute and beyond. Additionally, the ORACLE had on line a cathode ray tube plotter, a recorder, and a typewriter. Input and output was via punched paper tape. A high-speed Ferranti photoelectric reader was used for input. A slow paper tape punch was used for output. There were three thousand vacuum tubes in the ORACLE.

The ORACLE had one great redeeming quality. It was the fastest computer in the world at the time of its completion. Multiplication time was 240 microseconds (a microsecond is one millionth of a second) and division time was 440 microseconds These were phenomenal speeds in the early 1950s. There were over one hundred program instructions on the ORACLE.

The ORACLE was the first significant computer installed in the South and attracted attention from universities and research agencies from across the country. Several universities sent faculty members and graduate students to Oak Ridge to spend a semester. Bob Mann, a faculty member from the University of North Carolina, arrived at about the same time I moved to Oak Ridge. Bob was a professor in applied mathematics, and there were not many of them around in the early 1950s.

Bob and I were assigned as a team to learn the ORACLE. He was to be the analyst, and I was to be the programmer. We decided to write a program to calculate the roots of a fourth-degree polynomial. This was a problem we could solve analytically to check the calculated answers.

Bob and I did the calculations for a fourth-degree polynomial of the form

$$x^4+Ax^3+Bx^2+Cx+D=0$$

where A, B, C, and D are any real numbers. As students of algebra know, this problem is solved by finding the values of x which will satisfy the equation, (i.e. will make the left side equal zero for the given values of A, B, C, and D).

The computer program to do this will be iterative as opposed to the step-by-step procedure used to solve the differential equations to establish trajectories. The procedure first guesses at a solution then continually improves this by adjusting the previous guess. The procedure continues until the error at the end is less than a predetermined accuracy requirement. A successful procedure is one that provides a good adjustment to the previous result and provides quick convergence. A fourth-degree polynomial may have four, two, or zero real roots.

The computer program to find the roots is straightforward. For a given *A*, *B*, *C*, and *D*, one makes a guess at an *x*, evaluates the polynomial and looks at the difference between the result and zero. Then the correction (there are different schemes in use) is applied, and the process is repeated until the convergence criteria are met. This may require going through this iteration dozens of times, but this is what a computer does best. A big problem with programming the ORACLE was the scaling problem, making sure every calculation ended with a result between -1 and +1.

Bob made the calculations, and I prepared the program. It had to be punched into paper tape on a teletypewriter similar to those used by Western Union. In a couple of days, we had our program ready to go.

We scheduled our few minutes on the ORACLE and waited in line. The operator called us into the room and put our paper tape under the reader. We told the operator where to transfer control to pick up the first instruction. He did, and the computer gave what sounded like an ugly burp. We picked up our paper tape and slipped out of the room, disappointed that the program didn't run. I spent the next twenty-four hours going over every instruction and could not find a bug.

The next day we tried again, hoping the computer had made the mistake on the previous day. Again we got the same ugly burp.

As we again folded our papers in shame, the operator suggested we look at the punched paper-tape output, which

Left to right: Chuck Hubbard and Dr. Alston Householder with the ORACLE, 1953.

happened when we thought the computer had burped. We rushed to print the output on the teletypewriter and found the four roots correct to six decimals as Bob had calculated!

The speed of the ORACLE had flabbergasted us. We were accustomed to a CPC which would have taken a few minutes to get these results. I had, for probably the only time in my fifty years of computing, written a perfect program on the first pass.

It could not have happened at a worse time!

When I had been at Oak Ridge for a couple of months and had finished a period of orientation, a meeting was called in which Dr. Householder, Bob Mann, Dr. P. R. Kasten (a distinguished nuclear physicist), and I were instructed to develop a computer program which would study the requirements for a nuclear reactor to become critical—and stay critical—given a specific geometry and material makeup. This would involve designing a reactor which produced as many neutrons as are being absorbed; a condition called a steady state. The program we were to develop was called a nuclear code and was to be one of the first to be done in two dimensions (giving projected results in three dimensions).

Dr. Kasten was to establish the mathematical equations governing the process and provide other facts about a nuclear reaction, Dr. Householder and Bob were to set the guidelines for the numerical procedure, and I was to program it for the ORACLE. It was easy to see that this was to be a major programming task to take several months and would require a complete knowledge of the capability of the ORACLE.

I left the room and started looking for Chuck Hubbard.

CHAPTER 8

CHUCK HUBBARD: A GOOD BUDDY

I had concluded that Chuck Hubbard knew more about programming the ORACLE than anyone, and he fully understood the electronics of the computer and its capabilities. Chuck had joined Dr. Householder's group a year or so earlier than I, and was immediately given the job of chief programmer for the ORACLE. He subsequently wrote all the diagnostic routines to check the performance of each of the computer components.

Chuck came to Oak Ridge in 1952 after completing a master's degree in mathematics at the University of Illinois, where he had become acquainted with the ILLIAC, a cousin and predecessor of the ORACLE, which was then under development by the university. Since the ILLIAC was not yet completed, the programming was done from a set of handwritten notes the professor had prepared. This crude and rudimentary introduction to computing led Chuck to the decision that he would spend his life working with computers.

Chuck Hubbard was born in Iowa in the mid-1920s of parents who had been reared on farms. His father had left the family farm

In January 1948, the University of Illinois had appropriated $150,000 for the construction of a computer by the Reeves Instrument Company of New York; it was to be the first computer installed on the campus of an American university.

As the completion of the computer took longer than expected, John von Neumann, who was serving as a consultant to the university, suggested the university build its own computer.

Permission was granted to the university by the Institute for Advanced Studies (IAS) to build a copy of the ORDVAC computer which was being built by the IAS for the Aberdeen Proving Ground.

The ORDVAC was completed in February 1952, and the ILLIAC was completed and delivered in September 1952, becoming the first electronic computer built and completely owned by a university. The computer could store 512 forty-bit words (later expanded to 1,024) in main memory and had a magnetic drum auxiliary storage with a capacity of 12,800 words.

Over the years, the ILLIAC was updated and replaced with newer versions. While most of the early one-of-a-kind computers were replaced by those being offered by computer manufacturers, the ILLIAC remained and was improved, while testing new and novel computer advances such as parallel programming and different type computer memories.

In 1955, a four-bit transistorized computer was built and tested by the Digital Computer Laboratory at the University of Illinois and, in 1957, the faculty demonstrated a flip-flop (which determines computer speed) ten times faster than any other design then in use.

In 1964, the university, with sponsorship and funding from the air force, took a major step in contracting with the Burroughs Corporation to build ILLIAC IV. This computer

> was to be the most powerful computer in the world and was to have many new and innovative features. However, the project was met with considerable hostility by campus protestors who were angry that the university was in close association with the military on the development project. In May 1969, the boiling point was reached; to avoid violence, the project was moved to NASA's Ames Research Center at Moffett Field in California. The project, under the supervision of D. L. Slotnick, was completed with considerable success in 1972.

and started a restaurant. It was successful, and he soon had three restaurants in small towns in central Iowa.

Things went well for the family until the big stock market crash of 1929 and the resulting Depression. Mr. Hubbard lost his restaurants but managed to pay off everyone he owed; bankruptcy was not an option ever considered. Chuck's father accepted a job with the postal service in Chicago; and Chuck grew up in a nearby suburb.

Chuck remembered being very proud when his father was once asked who they got to repair things around the house and his father replied that he could fix everything except the radio, and Sonny (Chuck) could fix that.

When Chuck completed a BS degree in mathematics at the University of Miami in Ohio in 1950, he accepted a job with the Navy Electronics Laboratory (NEL) in San Diego. After a year, he decided he needed further training if he was to excel—and for Chuck, anything short of understanding it all would not be satisfactory. He realized that he needed more training in higher mathematics, so he entered the University of Illinois in the fall of 1951. He received a master's degree in 1952.

Visiting his parents in their retirement home in a settlement called the Homestead near Crossville, Tennessee, Chuck learned of the computer development at Oak Ridge. He visited with Dr. Householder in the guard shack at ORNL and made application for a job. Soon after returning to NEL, Chuck received a job offer and joined the math panel in 1952. He spent a considerable amount of time at the Argonne National Laboratory as work was completed on the ORACLE.

Chuck and I became lifelong friends. He agreed to tell me everything he knew about the ORACLE, and I agreed to teach him how to bass fish. He had never caught a fish. We soon fixed that, since Norris Lake, the first of the TVA dams, and Watts Bar Lake were nearby.

Chapter 9

An Early Nuclear Computer Code

The attempt to understand the basic properties of nuclear chain reactors through a detailed description of the behavior of the neutrons which support the reaction is known as reactor analysis. The mathematical equations which give an accurate process to count the neutrons can be summed up in a set of partial differential equations named the equations of the Boltzmann type, best known for their use in the kinetic theory of gases. The equations become intractable when one tries to derive rigorous predictions of the behavior of neutrons in cases of practical interest; thus reactor analysis quickly turns to approximations which yield useful results and whose validity is judged by comparison with experiments.

A nuclear chain reaction is caused when a neutron is captured by a heavy atomic nucleus which then splits, yielding energy—the useful by-product obeying Einstein's statement of the relation between energy and mass—and a few neutrons, some of which will cause other splits, thus resulting in a chain reaction.

There are several species of atoms where fission may take place and the one most commonly used in the 1950s was U-235. The number of neutrons produced with each neutron split varies but was determined by experiment to average 2.5 neutrons per split for U-235.

The focus of attention in reactor analysis is on the neutrons released by the fission, and how many will disappear and how many will go on to cause other fissions. The analysis consists of tracing the life histories of the released neutrons, of their daughters, and of all subsequent generations. This study depends on the knowledge of the materials and geometry of the assembly in which the neutrons are moving.

Nuclear reactors have the following general structure:

- A central region called a *core* (containing the fuel, U-235 for example).
- A shell surrounding the core called a *reflector* and *moderator* (graphite or heavy water).
- An outer shell called the *shield* (lead, concrete).
- *Control rods* to absorb neutrons and control the reaction (cadmium).

The reflector contains material which moderates (slows) the neutrons and cools the core. "Moderator" also means that the speed of some neutrons will be slowed, thus giving a better chance of fission. Fast neutrons sometimes slide right past atoms and produce no fission. A slow or wobbly neutron, traveling at a speed slightly less than the speed of light, has a better chance of accomplishing fission (in uranium-235). The shield serves as a stopping medium for dangerous radiation from the core.

When a neutron is born from fission in the core, it may wander around in the core or pass into the reflector and back to the core until it is absorbed by some nucleus or passes out of the system. The entire lifetime of a neutron in the reactor may be in the order of one one-thousandth of a second to one one-ten thousandth of a second; the neutrons are moving at speeds near the speed of light.

When a neutron is absorbed by another nucleus there is a chance for fission and the production of more neutrons. If there is to be a stable chain reaction, the neutrons from a given fission must, on the average, produce one neutron which causes fission. The control rods serve as an absorber of neutrons if too many are produced and there is a need to slow the reaction.

Stated mathematically, a steady state exists and a reactor is critical if

$$A \times F = 1.$$

If $A \times F < 1$ the fission dies. If $A \times F > 1$ then there is a possible nuclear problem.

Where A = average number of neutrons produced per fission process.

Where F = fraction of neutrons which eventually cause a fission in the fuel.

The numerical procedure to do a reactor analysis is iterative and is the process of searching for the materials and geometry which cause a convergence satisfying the formula above. In solving a set of ordinary differential equations, such as those which define the trajectory of a missile, the procedure is step-by-step. In the case of solving a set of partial differential equations, describing a two-dimensional problem, the procedure is iterative.

At Oak Ridge, the nuclear code we had decided to develop was a three-region (material types) with two groups (fast and slow neutrons), and we would use cylindrical coordinates.

The first step in solving any physical problem, whether it is rockets or reactors, is to develop the set of mathematical equations which define the process. The second step is to solve the set of equations, given a set of boundary, or initial, conditions. Since in almost every case a closed form (analytic) solution is impossible, one must develop a numerical, step-by-step or iterative, approach to reach a solution. The mathematical representation of the reactor problem is a set of second-order partial differential equations with an extensive set of boundary conditions. The set of differential equations is replaced by a set of difference equations (an algebraic set) which may, in this case, be depicted as a two-dimensional mesh laid down over the geometry of the reactor. The unknowns in the algebraic system are the mesh points (corners of each mesh square) which form a large square matrix (determinant). The numerical procedure then involves the solution of two very large systems of algebraic equations, where the method of solution is iterative, meaning the system will be evaluated many times until convergence is reached.

There are two groups of neutrons, fast and slow, and it is necessary to follow each group. The solution to these systems of algebraic equations will allow one to approximate the production and loss of neutrons within the reactor, and this result will allow one to compute the effective multiplication constant k_{eff}, or simply K, for the reactor.

If K is less than 1, the reactor is subcritical and the reaction will die out. If K is greater than one, then the reactor is called supercritical and may become a nuclear problem. One is searching for the materials and the geometry which will result in a K equal to 1 which is called critical and then the reactor will set about to produce energy. (Note: The exhibits which follow give the geometry and the equations which are used in determining K).

In solving a numerical problem such as the one described above, there is always a search for ways to accelerate convergence to the end result. In addition to his work in estimating errors in numerical procedures, Dr. Householder had become expert in developing procedures to accelerate convergence.

SUMMARY

The code described in this report has served as an intermediate step to the multi-region, multi-group, two-dimensional reactor calculation. The particular geometry of the code was chosen since there was an immediate need for results from a code of this type. The problem solved is, in general, a three region cylindrical problem using two groups of neutrons. Provision is also made for a thin shell surrounding the core. This shell is considered to be transparent to fast neutrons and an absorber of thermal neutrons. The code can be run with or without the shell and as a two-region problem if so desired. The code has been made as general as possible so that it may be applied to any problem which can be suitably approximated by a two-group three-region cylindrical model.

MATHEMATICAL DESCRIPTION OF THE PROBLEM

The problem solved by the code is indicated by the following figure, differential equations and boundary conditions:

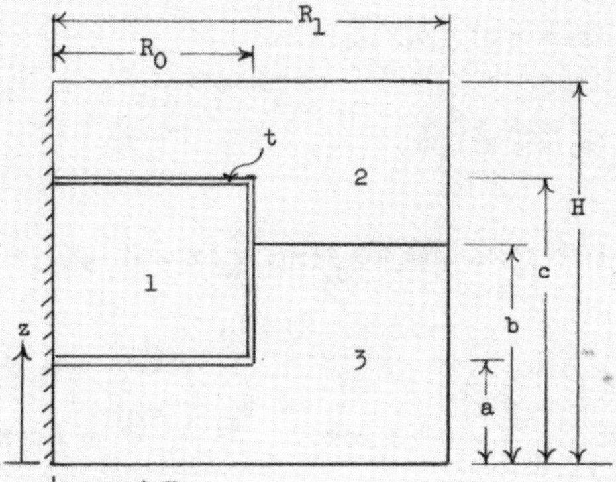

FIGURE 1

The geometry and summary description of the reactor code programmed for the ORACLE in 1954–55. From ORNL-2047 Physics 1956, *by C. L. Bradshaw.*

The figure indicates that cylindrical coordinates (r , z) are used and that symmetry is assumed about the z-axis.

Differential equations:

(1) $$D_{f_i} \nabla^2 \varphi_{f_i} + \frac{k_i}{p_i} \Sigma_{s_i} \varphi_{s_i} - \Sigma_{f_i} \varphi_{f_i} = 0$$

(2) $$D_{s_i} \nabla^2 \varphi_{s_i} + p_i \Sigma_{f_i} \varphi_{f_i} - \Sigma_{s_i} \varphi_{s_i} = 0$$

$i = 1, 2, 3$ refers to the three regions, s and f to the two neutron groups, slow and fast.

Boundary conditions:

(3.a) $$\left(\frac{\partial \varphi_{j_i}}{\partial r} \right)_{r=0} = 0 \quad ; \qquad i=1,2,3; \; j=f,s$$

(3.b) $$\varphi_{j_i}(r = R_1, z) = 0 \quad ; \qquad i=2,3 \; ; \; j=f, s$$

(3.c) $$\varphi_{j_3}(r, z = 0) = 0 \quad ; \qquad j = f, s \; .$$

(3.d) $$\varphi_{j_2}(r, z = H) = 0 \quad ; \qquad j = f, s \; .$$

(3.e) $$\varphi_{j_2}(r > R_o, z = b) = \varphi_{j_3}(r > R_o, z = b) \quad ; \qquad j = f, s \; .$$

(3.f) $$D_{j_2} \left(\frac{\partial \varphi_{j_2}}{\partial z} \right)_{r > R_o, z=b} = D_{j_3} \left(\frac{\partial \varphi_{j_3}}{\partial z} \right)_{r > R_o, z=b} \quad ; \; j = f, s$$

Above and opposite page: The system of partial differential equations and boundary conditions used to define the nuclear reactor code in 1954–55.

(3.g) $$\varphi_{j_2}(r < R_o,\ z = c) = \varphi_{j_1}(r < R_o,\ z = c) \quad ; \qquad j = f,\ s\ .$$

(3.h) $$D_{f_2}\left(\frac{\partial\varphi_{f_2}}{\partial z}\right)_{r < R_o, z=c} = D_{f_1}\left(\frac{\partial\varphi_{f_1}}{\partial z}\right)_{r < R_o,\ z=c}$$

(3.i) $$D_{s_2}\left(\frac{\partial\varphi_{s_2}}{\partial z}\right)_{r < R_o, z=c} = D_{s_1}\left(\frac{\partial\varphi_{s_1}}{\partial z}\right)_{r < R_o, z=c} + \Sigma_a\, t\, \varphi_{s_1}(r < R_o,\ z = c)$$

(3.j) $$\varphi_{j_3}(r < R_o,\ z = a) = \varphi_{j_1}(r < R_o,\ z = a) \quad ; \qquad j = f,\ s$$

(3.k) $$D_{f_3}\left(\frac{\partial\varphi_{f_3}}{\partial z}\right)_{r < R_o,\ z=a} = D_{f_1}\left(\frac{\partial\varphi_{f_1}}{\partial z}\right)_{r < R_o,\ z = a}$$

(3.ℓ) $$D_{s_3}\left(\frac{\partial\varphi_{s_3}}{\partial z}\right)_{r < R_o, z=a} = D_{s_1}\left(\frac{\partial\varphi_{s_1}}{\partial z}\right)_{r < R_o, z=a} - \Sigma_a\, t\, \varphi_{s_1}(r < R_o, z=a)$$

(3.m) $$\varphi_{j_1}(R_o,\ b < z < c) = \varphi_{j_2}(R_o,\ b < z < c) \quad ; \qquad j = f,\ s$$

(3.n) $$\varphi_{j_1}(R_o,\ a < z < b) = \varphi_{j_3}(R_o,\ a < z < b) \quad ; \qquad j = f,\ s$$

(3.o) $$D_{f_1}\left(\frac{\partial\varphi_{f_1}}{\partial r}\right)_{r=R_o,\, b < z < c} = D_{f_2}\left(\frac{\partial\varphi_{f_2}}{\partial r}\right)_{r = R_o,\, b < z < c}$$

(3.p) $$D_{f_1}\left(\frac{\partial\varphi_{f_1}}{\partial r}\right)_{r=R_o,\, a < z < b} = D_{f_3}\left(\frac{\partial\varphi_{f_3}}{\partial r}\right)_{r=R_o,\, a < z < b}$$

(3.q) $$-D_{s_1}\left(\frac{\partial\varphi_{s_1}}{\partial r}\right)_{R_o,\, b < z < c} = D_{s_2}\left(\frac{\partial\varphi_{s_2}}{\partial r}\right)_{R_o,\, b < z < c} + \Sigma_a\, t\, \varphi_{s_1}(R_o,\, b < z < c)$$

(3.r) $$-D_{s_1}\left(\frac{\partial\varphi_{s_1}}{\partial r}\right)_{R_o,\, a < z < b} = -D_{s_3}\left(\frac{\partial\varphi_{s_3}}{\partial r}\right)_{R_o,\, a < z < b} + \Sigma_a\, t\, \varphi_{s_1}(R_o,\, a < z < b$$

METHOD OF SOLUTION

It is assumed that the material make-up of the three regions is known and that one is interested in finding the effective multiplication constant for the given reactor system. In order to produce this number k_{eff}, (which will in all subsequent references be called simply K) , the set of differential equations is replaced by a finite difference system and this difference system is solved numerically. The numerical method requires the solution of two large systems of algebraic equations (one for each of the two groups) several times to gain the required convergence to K . An outline of the determination of K is as follows:

A. The differential system is replaced by a difference system using approximations of the form,

Boundary conditions and a brief summary of the method of solution for the nuclear code of 1954–55.

During the programming and running of the program described above, Dr. Householder and I had several discussions of techniques which might be used to accelerate convergence. After one of these discussions, Dr. Householder came back to me in two days with some fifteen handwritten sheets to suggest a method which he felt might accelerate convergence.

With the help of two or three colleagues, I was able to understand the proposed procedure in about two weeks. The procedure was elegant and contained some beautiful mathematics. If applied, it would surely accelerate convergence and give good results. However, to apply the method would require weeks of programming and was almost as complicated as my original problem, so I decided my original problem was enough of a challenge for me at the time.

The mean time between failures (MTBF) for the ORACLE was measured in just a few hours and it was unusual to have a twenty-four-hour period without some component failure. The vulnerable components were the vacuum tubes and the eighty large cathode-ray tubes used for memory.

A basic part of any computer program written for the ORACLE was a restart procedure. This procedure, which could be a cumbersome part of the program, required that the program and intermediate results be stored on magnetic tape at designated intervals, usually every ten minutes of computer time. This restart procedure was to assure the programmer did not lose more than ten minutes of compute time in case of failure of a computer component.

If it was determined that the Williams tubes were all working well, the ORACLE could run with 2,048 words of memory. The engineers would run diagnostic routines to keep track of the condition of all the tubes. If, on a certain day, one memory location was troublesome while running on 2,048 words, a bulletin would be issued to programmers to avoid using that particular location until further notice. The programmer would be required to make adjustments to avoid troublesome memory locations.

During the day shift, only short debugging or production runs were allowed, and these were usually limited to a few minutes. For large programs, the only satisfactory procedure was to get a four-hour block of time in the middle of the night and to become completely involved with the computer and the program. After a few nights of debugging, a person could run through the program in their mind, even in their sleep. During these late-night debugging periods, a programmer could get to know the night operator and the night janitor as well as their program. In many instances, the operator chose to get a few hours sleep, and the janitor would come in to keep the programmer company and to marvel at the flashing lights.

The program developed to meet the requirements of the nuclear scientists took several months to complete. The computer time required to run a simple geometry took a little more than one hour. The program represented an intermediate step to computer programs which accurately predict the material makeup and the amount of fuel to build a nuclear reactor with a required energy output.

Chapter 10

Searching for a PhD

The Oak Ridge National Laboratory (ORNL) was primarily a research laboratory and operated much as a research university. Promotion, to a large extent, was based on the quality and quantity of research and the publications resulting from that research. A PhD became a necessary tool to advance in the organization.

Chuck Hubbard and I each had a master of arts degree in mathematics and were pursuing PhDs in the evenings at the University of Tennessee. All of our classes were evening classes and only a few of them were offered in Oak Ridge; the others were offered on the campus in Knoxville. The classes in Knoxville required a fifty-mile drive twice a week, over a road which was not in the best of condition, and through some harsh East Tennessee rains and sleet.

In the 1940s and 1950s, all graduate mathematics consisted of pure mathematics—which meant one could not possibly apply it to a practical problem. The process of applying mathematics to a practical down-to-earth problem was called applied mathematics, and the prevailing attitude was that no good mathematician would spoil his reputation or soil his hands trying to solve a practical problem. Those problems should be left to the engineers, physicists, and perhaps the astronomers.

Chuck and I were taking a course in higher algebra, which was the purest of the pure. On this particular night in February 1956, the subject was something like "A fundamental theorem of non-Abelian semi-groups embedded in a Galois Field and invariant under all transformations of the second kind." We always took a lot of notes in class and then went over them together at a later time. When Chuck laid down his pencil and stopped taking notes during the first part of the lecture, I knew this was a significant moment in our lives. As it turned out, it steered our careers in an entirely different direction.

At the break in the class, Chuck carried his books out of the classroom, and I followed suit. He suggested we just go home instead of remaining for the rest of the lecture.

As we drove back to Oak Ridge, we both agreed that the courses we were taking were never going to do us any good unless we planned to teach at some university. I had decided against that when I left the teaching job at Tennessee Tech.

He also told me he was ready to leave Oak Ridge, as he had been there four years and would like to look for another challenge. I had been there only two and one-half years and still had some things to do. As we discussed other job opportunities in computing, we realized there were not many computer facilities in the country at the time.

My good friend Grady Holt had left Tennessee Tech a couple of years after my departure and was now recruiting technical personnel for Consolidated Vultee Aircraft in Fort Worth, Texas. I suggested to Chuck we might call Grady to see if they had computer facilities and needed programmers.

The next day I talked with Grady, and in fact, Consolidated Vultee Aircraft (later called Convair, a part of General Dynamics) had a good computer center started and was in need of experienced programmers. The computing group was headed by Hank Wolanski, an aerophysics engineer. We set a date for our visit, and Grady would make the proper arrangements to get Chuck application forms to precede our visit.

We set about scheduling our trip to Fort Worth. Chuck and I decided we would spend the first night in Huntsville, Alabama, since this was on the route we planned to take and we could get in a short visit with Doc Hoelzer and my other friends and still meet the travel schedule we had planned.

Doc had moved his computer center to a larger building and had expanded his staff to some forty people. The group exhibited a lot of energy which emanated from Doc and his principal aide, a young army lieutenant from the West Coast named Tac Yamashita who was in charge of getting Doc the resources he needed to continue to build his computer center. Right away, it crossed my mind that it was ironic we were seeing the computer facilities to support our most important military project managed by a German and a Japanese even though the war had been over for barely a decade.

When I introduced Chuck to Doc and told him a little of his background, I saw a gleam in Doc's eyes. He latched onto Chuck and proceeded to ignore me. My feelings could have been hurt, but I quickly guessed what Doc had in mind. He took Chuck into his conference room, where he and Yamashita put on a full-court press. I was left in the hall to spend time with other friends.

During the next three hours, Doc dragged every employee in his center through the conference room and introduced them to Chuck. I was certain he had received Doc's pencil demonstration of flight. Doc probably developed the equations of flight and control, and told Chuck it was his patriotic duty to help them solve the computing problems necessary to build the rockets to protect our country.

It was almost noon before Chuck was released, and as we left the building there was a stunned look on his face and his eyes were glazed over. He looked like a bird which had just flown into a transparent window and was still trying to figure out what had happened. As we headed to the car, I explained our need to hurry along so we could meet our commitments to Grady and Consolidated Vultee.

As we left Huntsville and headed west, Chuck was still trying to recover. He looked over at me and said he didn't think there was any need to hurry on to Fort Worth. When I inquired why, he replied rather limply, "I think I just accepted a job in Huntsville."

We proceeded on to Fort Worth on a more or less pleasure trip to visit Grady, look at the facilities at Consolidated Vultee, and meet with Wolanski. Like everyone else, they had an IBM 650 on hand and were waiting on the next generation from IBM. We went through the paces of job application, although there was no enthusiasm; but we had a good visit with Grady and his family.

The trip back to Oak Ridge was highlighted by a wonderful lobster dinner at a restaurant in Texarkana where the owner sat down and became acquainted with all her customers. In the days before interstates, traveling was much more of a challenge and more of a good experience. The little filling stations at grocery stores along the way, where bologna sandwiches sold for fifteen cents, gave you a good opportunity to meet the local folks who always asked where you had been (and why) and where you were going (and why). They would always tell you about problems in the road and where to watch for speedtraps.

We were driving Chuck's relatively new red Buick Century, and we assumed we might be a good target for the local sheriffs. During the trip, we discussed whether or not we could outrun the jalopies we had seen some of the deputies driving, but decided we would not try since communications was now coming to even remote areas, and they could set up a roadblock.

I continued to work on and try to improve the computer time required to do the reactor analysis using the program I had developed. Most of the computing time was spent in solving the large systems of algebraic equations over and over again until the convergence criteria were met. At the end of each pass in the iterative process, a set of factors was used to accelerate convergence. These factors were usually called relaxation (or convergence) factors, and it was fortunate that Dr. Householder was probably the world's foremost expert in convergence acceleration and error analysis in computing processes.

One day while consulting with Bob Mann, the security guard from the main gate came to my door with a telegram. I had received very few telegrams in my life and certainly none at my office, so I was sure it was not good news. I read the telegram which basically read as follows: "The Army Ballistic Missile Agency is pleased to offer you the position of Mathematician GS-13 in the Computation Laboratory of the Army Ballistic Missile Agency (ABMA). Please inform us at your earliest convenience of the date you will be reporting for work."

This had to be a mistake since I had not applied for a job. The telegram was meant for Chuck, and Doc Hoelzer had made a big mistake. What a joke on him.

I hurried to Chuck's office with the telegram, and when I entered, he had just completed reading a telegram just like mine—except it was addressed to him.

Stunned, I realized that the joke was on me and Doc had pulled a fast one. He knew if he could get Chuck to Huntsville, I would come along. We started making plans for the move to Huntsville.

Chapter 11

Army Ballistic Missile Agency

In 1955 the hot war in Korea ended and the cold war with Russia (the USSR) became more intense. The Redstone missile neared completion and was readied for deployment as the first missile battery of the United States. Several intelligence sources reported that the USSR was advancing faster than the United States in missile technology, and some believed that the USSR had an Intermediate Range Ballistic Missile (IRBM) well under way. This missile would have a range of fifteen hundred to two thousand miles, making all the major cities of Central Europe possible targets. This was a cause for concern and gave emphasis to plans by the United States to take the next step after the Redstone (two-hundred-mile range) and move rapidly to develop an IRBM (fifteen-hundred-mile range) of its own.

The United States Defense Department went into a crash program to develop the IRBM and instructed the army, air force, and navy to proceed with individual development plans to build their versions of this enhanced missile. A decision would be made at a later time as to which of the systems would go into production and be deployed. The army missile was known as the Jupiter and the air force missile was known as the Thor. The air force selected the Douglas Aircraft Company in California as the prime contractor to develop the Thor missile; the army selected

Dr. von Braun's group to develop the Jupiter; and the navy decided to join with the army in the development of the Jupiter and later modify the Jupiter to meet the needs of the navy.

The arrangement between the army and the navy was dissolved in less than a year. The navy very soon recognized that the missile to be carried aboard the submarine would need to use solid propellants and be placed on the submarine fueled and ready to launch. The Jupiter was to be a larger Redstone that would use liquid propellants and be fueled just prior to launch. The navy was given authority to develop an IRBM, the Polaris, to be used with the submarine fleet; the race was on to develop the first IRBM.

In order to give the IRBM project the highest priority, the army formed a new agency to be located on Redstone Arsenal in Huntsville and named it the Army Ballistic Missile Agency (ABMA). The army assigned one of its most dynamic officers, Major General John Bruce Medaris, to head the agency. The Medaris–von Braun team proved to be a good arrangement.

Chuck and I reported to work at ABMA on a Monday morning. Doc assigned Chuck to an office close to the computer room since he was to be head of systems and operation. I was to be head of a programming group called Special Projects.

Doc personally led me to my "office"—which was the third booth in the men's rest room! A small desk and telephone had been arranged in the cubicle, and my nameplate was properly placed on the door. Doc brought along most of his staff to welcome me. The next twenty minutes was spent joshing and crowding in to see my cubicle.

Doc's sense of humor was always evident, even in the most serious moments. A few years later when I served as his deputy director, we lost funding for our plans for a very important computer project. As I sat in my office lamenting the loss, Doc came in and told me a couple of jokes to cheer me up. I asked Doc how he could joke at a time like that when we had just lost a major project. He replied, and I will always remember it, "Brad, if you don't have fun at a funeral, nobody will come."

Doc was also a practical joker, but was always careful not to hurt feelings or play on personal shortcomings. Chuck and I both bought Volkswagens a few months after the move to Huntsville. Chuck spent the weekends tuning the engine on his car, and on Mondays he would tell us how many miles to the gallon he was getting.

Doc asked a couple of us to bring some gasoline to work, and we would add a quart to Chuck's car every day or so. He kept tuning, and the gas mileage kept increasing; up to thirty-five miles to the gallon. My Volkswagen was getting only twenty-three miles to the gallon, and Chuck was declared by all to be an automotive genius.

Then Doc stopped putting the gas in and started taking some out. The gas mileage dropped to twenty-five, then to fifteen miles to the gallon. Chuck was in a panic until he discovered the shenanigans, and we all had a big laugh.

The computation laboratory had grown to about forty people and the digital computing group was organized into three branches. Chuck and I were assigned as branch chiefs. The third branch chief was Carl Prince, who had a forceful personality and had been a chief boatswain's mate in the navy. Carl let us know right away that it was his ship, and we need not try to take over; he would consider it an act of mutiny. We assured him that there would be enough work for all of us, and over the years the three of us became good, close friends and were able to fight it out without lingering hard feelings.

There were two IBM 650 computers installed and an IBM 704 on order. I learned as much as I could about the 650 while preparing for the 704.

The 650 put IBM in the computer business for good, as hundreds of them were sold (leased), and it was said to be the first computer to make a profit for its manufacturer. The IBM 650 used a drum memory organized into signed ten-digit decimal words, and was unique in that it was a decimal machine just when everyone had decided that the electronic digital computer should use binary arithmetic. The basic system had two thousand ten-digit words of memory and was designed to be cheap (*only* a half-million dollars!) and easy to use. Not many people thought it was easy to use, and none thought it was cheap.

There were two accumulators, and the programmer used one or the other depending on the nature of the program step. If the program was stored sequentially on the drum, then each next instruction would need to wait a full drum revolution to come under the read heads. What one needed to do was to store the program on the drum in such a manner that the next instruction would be available just when the previous instruction was completed. This was called optimization of the program and required considerable skill and thought.

The difficulties in laying out the program on the drum for optimization led to the development of a very good assembly and optimization program. The program was called the Symbolic Optimization Assembly Program (SOAP) and was developed by Dr. Alan Perlis and associates at Purdue University. SOAP set the standard for assembly programs which followed. Dr. Perlis was a leading computer scientist of the day who later became head of the department of computer science at Yale University and was one of the early presidents of the Association for Computing Machinery (ACM).

During our first months at ABMA, the navy accelerated its Polaris missile program and placed it under the management of Adm. Hyman Rickover. It became one of the best managed programs in naval history, and no doubt, later became our most effective deterrent during the cold war with the USSR. The Polaris program also introduced the computer to the management process by developing a program called Program Evaluation and Review Technique (PERT). Over the years, almost every major construction and engineering project has been controlled and expedited by a Project Management System (PMS) similar to PERT.

The IBM 704 scheduled for ABMA was being built in Poughkeepsie, New York. At IBM, as an order was received, a computer on the line was scheduled for delivery and a large sign was placed on it to designate who would be the eventual owner. The complete trip of the computer down the assembly line took several months.

I went to Poughkeepsie to see our system and to see the men and women put it together. There was neither automation nor robots. Everything was done manually and with great pride by IBM employees. In one area, women threaded tiny wires through small, doughnut-shaped magnets to build magnetic core planes to serve as the main memory on the 704s. Each employee gladly stopped to tell me what they were doing and how important his or her work was to the successful completion of the computer.

The IBM 704 was a reliable computer and was easy to program. It had some twenty-four easy-to-understand instructions; 4,096 thirty-six-bit words (expandable to 32K words); operated in floating point as well as fixed point; had index registers to count passes through a loop; and would operate for hours, even a few days, without a hardware failure. The line printer attached to the computer allowed one to dump the entire memory in just a few minutes. Input for program and data was by punched cards, and each computer installation needed a basic set of card equipment to allow program and data preparation.

During this period, there was considerable stress and growing pains in the three branches of the armed forces. At the end of World War II, Congress and President Truman formed the Department of Defense and named James Forrestal as its first secretary. For the first time, the three military services had a common boss (other than the president). The air force, created from the army air corps, became a full member of the Defense Department and struggled to gain acceptance and equal status, and a new air force academy was being built at Colorado Springs, Colorado.

The development of the guided missile was bringing about a need for a clarification of the roles and missions of each of the services in the development of this new weapon system. The army was making more progress in missile development because of its good fortune of having the von Braun group.

The Defense Department, in concert with President Eisenhower, made a decision on roles and missions which still stands today and has evidently served the country well. The navy was given the mission to develop all guided missiles to be placed aboard and fired from ships at sea. This, of course, gave the navy the Polaris submarine missile system and smaller missiles to be launched from shipboard.

The air force, which already had the mission of long-range bombing, was given responsibility for the development and deployment of Intercontinental Ballistic Missiles (ICBM).

The army was restricted to the development and deployment of missiles of a range less than three hundred miles, which was equated to long-range cannon and similar air bombardment. This ruling was a disappointment for the von Braun group, which had hoped to continue developing larger missiles. The ruling also meant that the developed Jupiter would be turned over to the air force for deployment, and one might assume the Thor would get favored treatment. The army, which historically had the mission of short-range bombardment and the development of anti-aircraft weapons, retained the same mission in the missile age. This latter mission has led the army into the major role in antimissile development and deployment (including the Strategic Defense Initiative, known as Star Wars).

The Jupiter missile was to have a range of twelve hundred to two thousand miles and an accuracy of less than five hundred yards. The ruling of the Defense Department concerning roles and mission did not alter the role of the army in the development of the Jupiter, only its deployment. As compared with the Redstone, which was tall and slender, the Jupiter was short and fat. The tanks were to carry considerably more fuel, and the missile was to deliver a heavier payload than the Redstone; a larger engine would be required. The thrust of the Redstone engine was fifty thousand pounds while the Jupiter engine would need a thrust of one hundred fifty thousand pounds. Companies such as General Electric, North American Aviation, and Pratt-Whitney were heavily involved in rocket engine development.

The Research Projects Laboratory of ABMA was headed by Dr. Ernst Stuhlinger, a von Braun associate for many years. One of his top assistants—and the principal astrophysicist—was Dr. Chuck Lundquist, a dedicated scientist with abundant energy who typically had a fifteen-hour workday. Lundquist came to Huntsville as a drafted GI from Penn State University and had decided to make his career there.

There was speculation that the United States would choose to place an artificial satellite in orbit as part of the celebration of the International Geophysical Year, which covered the years 1956–58. As it turned out, the United States and the USSR both decided to place artificial satellites in orbit during that period. The Research Projects Laboratory had hopes that ABMA would be assigned the mission to put the United States satellite into orbit.

The Computation Laboratory became involved with Chuck Lundquist in doing lifetime studies of satellites placed in orbit with varying configurations and initial conditions on injection. The project Lundquist had under way was to determine how long (lifetime) a satellite would stay in orbit with only the force of gravity and a small air resistance acting on the body. The calculations needed to track the path of an artificial earth satellite were based on physical laws established by the astronomer Johannes Kepler, rather than on the laws of Isaac Newton used to calculate trajectories of rockets across the surface of the earth.

The fundamental laws of physics determined that a near circular orbit, at two hundred miles above the earth's surface, required an injection velocity of approximately 17,500 miles per hour (mph). A slightly higher velocity would stretch the elliptic orbit until, at 25,000 mph, an escape velocity would be reached and the satellite would go into an orbit around the sun. A velocity of less than 17,500 mph would decrease the orbit and the lifetime. If much less than 17,000 mph, the satellite would fail to orbit and return to earth.

The farther from the earth, the less the injection velocity would become. If the body was as far away as the moon (240,000 miles), then two thousand miles per hour is all that would be required. The energy trade-off was, of course, the bigger the satellite and the farther from earth, the more energy was required to get to the point of injection.

These lifetime studies required the numerical solution (step-by-step) of a set of ordinary differential equations. The process was less complicated than that required in solving the equations defining the trajectory of a guided missile. However, there was one very complicating factor: the calculations for the orbiting satellite went on, and on, and on, giving rise to a problem not encountered before. This was the problem of the round-off error building up to a magnitude which impacted the significance and accuracy of the calculated numbers.

A round-off error occurs on each multiplication of two numbers, since the product of two eight-digit numbers must be rounded off in the ninth place, causing an inaccuracy of as much as one-half digit. If this is done a million times, the error can be significant.

The problem of round-off comes face to face with the other error caused by numerical integration: the truncation error. The truncation error is caused by the method used in doing a numerical integration, which is an approximate solution with an inherent error. The shorter the time steps were made, or the finer the mesh, the smaller the truncation error. However, the shorter the time step, the more computation must be done and the larger the round-off error.

Chuck Lundquist named this the "diabolical problem" of orbit computations. We later received help from astronomers who had been calculating the orbits of natural satellites all their lives. Professor Leland Cunningham of the University of California, Berkeley, was particularly helpful and gave good advice in rewriting the computer programs which allowed for the successful tracking of *Explorer 1*.

A young GI named David Freeman had been assigned to work with Lundquist. David was a recent graduate of Harvard and had taken several applied mathematics courses at MIT. He was a bright, dedicated young man and had twice been to the IBM test center in New York to check out his programs but he had not been successful.

As a computer was being built by IBM, each customer was given a block of free checkout time to be used on the computer in New York City. IBM had placed a 704 at 590 Madison Avenue with a large window so that persons on the sidewalk could view this marvel of electronics.

Doc called me into a private conference with Chuck Lundquist to see if I could help without appearing to butt into what Freeman was doing. David was working for fifty dollars a month (a private's salary). He was very sensitive and conscientious, and he was justly proud of his academic background.

I very tactfully asked David to allow me to review his work for my own education, and he was delighted go through his program with me. His program was written using a sophisticated difference scheme (for numerical integration) rather than a usual straightforward integration scheme with which I was familiar. I surmised that his problem might be that at a certain point in the calculation, the difference scheme became unstable, and the calculation went to nonsense.

David believed he had a bug in his program, and he kept looking for the bug instead of considering other possibilities. I decided that I would secretly write the program using a simpler yet completely stable numerical integration system called the Runge-Kutta method. I then went to New York with David to see if I could figure some way to ease this method into the calculations.

The IBM test center at 590 Madison Avenue was run like a well-disciplined ship. Each user was assigned a ten-minute slot during the day to do program checkout, and offices in the upper floors of the building were provided for use by customers. Customers were called a few minutes before their time slots and asked to report to the dispatcher, located on a glass-enclosed balcony constructed above the computer, and reminding me of the quarterdeck on a ship. The dispatcher could see all that was going on in the computer room and would dispatch a user down the stairway when the programmer's time slot became available.

The first try for David failed as it had before and he continued to look for a program error.

On the second day, I secretly punched my card deck and had it ready. When his method failed again, I very gently and apologetically asked if I could try something I had put together.

He seemed surprised and was sure I had not had time to do anything worthwhile. He did not know I had been working on this for several days.

We watched the calculations take place, and the results agreed very well with his results. When we came to the point where his program had been failing, David leaned over the printer and was amazed that the calculations continued. It was a victory of simplicity over sophistication.

David reluctantly agreed to take over my program and finish the series of calculations he had brought to New York. The method I had used was a crude method as compared with the elegant difference scheme used by David, and its success was due to the numerical procedure used and not to the superiority of the computer program.

A few weeks later when the computer was installed in Huntsville, we changed the program to include additional forces acting on the satellite and obtained different results. When a satellite is placed in orbit, the natural expectancy is that the satellite will gradually get closer to earth due to the pull of gravity, and finally burn up on reentry. The purpose of lifetime studies was to see how long the satellite would stay in orbit with a given set of injection conditions (velocity, orientation, and mass). This was not happening in our new program, as it projected the satellite would get a little further away from the earth on each orbit around the earth.

This was unusual and precipitated some good discussions with Chuck Lundquist, who discussed this with his boss, Dr. Stuhlinger—who found it interesting enough to talk it over with Dr. von Braun.

The three of them went through various scenarios to explain this strange phenomenon. The most logical conclusion was that there could be a slingshot effect as the satellite came out of its perigee (point closest to earth), which could be slightly enlarging the orbit.

This discussion was hot for about four days before I discovered a mistake in the computer program. There was a minus sign where there should have been a plus sign, and this was the complete cause of the remarkable phenomenon. The next time Dr. von Braun saw me, he kidded and laughed about it. David got a good, big smile from the incident, as if to say, "Gotcha back."

In addition to the expected problems of a larger engine with larger tanks and more complicated plumbing and guidance, two unexpected problems were encountered with the Jupiter. The fuel tanks, which were short and fat as compared with the Redstone, led to a problem of fuel sloshing as the fuel level became lower in the tanks. This sloshing led to an instability, which led to an uncontrollable missile.

This problem was solved through the installation of a system of well-engineered baffles in the fuel tanks. The computer program to design and place these baffles involved the solution of systems of partial differential equations, which could be approximated by evaluating many large determinants (square matrices).

The engineer from the Aeroballistics laboratory who was assigned the problem was named Phillip Wolfgang Zettler-Seidle. He was a brilliant young physicist and engineer who had joined the group directly from Germany as a result of the considerable recruiting effort in the early 1950s. Phillip was a good engineer but knew little about computers.

I had gone into a rush mode to write a computer program to evaluate large (up to thirty-by-thirty) determinants. This took three or four days and nights. We then went into all-night sessions to run the cases Phillip thought necessary to settle the question of where to place the baffles in the Jupiter tanks.

After a few nights of running various combinations of many situations, I thought we were done. Then Phillip announced we needed to run the same set of cases using complex numbers of the form $a+bi$.

The student of algebra will recall that in the complex number, a and b are real and i is the imaginary unit, the square root of -1. Generally the student first meets imaginary numbers when finding the roots of a second degree polynomial (binomial) using the quadratic formula and negative numbers are calculated under the radical.

The student of algebra will also recall that calculating the product of $(a+bi)$ and $(c+di)$ will take six arithmetic steps instead of the one for a real number. The process of dividing complex numbers is even worse, and each of these steps would need to be programmed.

As I expressed some dismay at this turn of events and told Phillip that this would require more days of programming, he was surprised and suggested that I simply turn on the switch so the computer would do complex arithmetic. This was typical of the presumed sophistication given to the early computers. They were, after all, called "giant brains." We took a break, cussed a little, had a good laugh and a cup of coffee, and set about the task of writing a computer program to evaluate determinants with complex coefficients.

The second, and more serious, problem with the Jupiter missile was the destruction of the warhead on reentry (from near outer space) due to the excessive heat generated as the warhead reentered the earth's atmosphere. It was determined that, in addition to computer models, it would be necessary to fly small warheads and recover them to examine the damage done on reentry. In order to test the heat shields and materials for the warhead, a Jupiter-C missile was constructed.

The Jupiter-C consisted of an elongated Redstone with an upper stage made up of a set of small solid-propellant rockets mounted on a rotating platform for stability. These small solid-propellant rockets would be fired at the proper time to hurl the model warhead into space for a reentry test as a model Jupiter warhead would be mounted on top of the upper stage with various material configurations. Several combinations of material including mylars and ceramics were selected for testing.

After reentry the model warhead was picked up downrange in the Atlantic Ocean by waiting planes and ships. The Jupiter-C experiments led to the solution of the reentry problem of the Jupiter but did not solve the general problem of reentry for all time. The problem of a reentering body from space is still an area requiring study.

In addition to the assigned mission of completing the Jupiter missile, a great deal of excitement was generated over the possibility of launching an earth satellite. The Jupiter-C could do the job with just minor modifications; however ABMA was to be bitterly disappointed as President Eisenhower and his cabinet made the decision that, since this was to be a scientific experiment, the U.S. would not use missiles or other devices built for a military purpose. The use of the Jupiter-C was not an acceptable option.

President Eisenhower made this decision in line with the feeling that the military should not become involved in this purely scientific endeavor, and the U.S. should exhibit to the world its hope for peaceful and cooperative projects.

The Naval Research Center at Dahlgren, Virginia, was given the mission, which would require the design and development of a whole new rocket system, known as Vanguard. President Eisenhower and his cabinet had made a serious mistake in judgment concerning the importance of this project to world leadership. Due to the inexperience of the Vanguard development group, the project was doomed to failure, and Russia would beat the U.S. into space using military rockets.

General Medaris and ABMA put up strong protests to the decision made by President Eisenhower. They insisted that a modified Jupiter-C could place a satellite in orbit within three months from an order to proceed. They made argument after argument, but to no avail.

An army colonel named John C. Nickerson was so upset with the decision that he released some classified information to Drew Pearson, a noted muck-raking journalist with a syndicated column. The information released by Nickerson was to the effect that the army had a missile ready to go that could place a satellite into orbit, but was being prohibited from using it or talking about it. (There was no national security reason for this information to be classified; it was classified for political reasons, as is so often done.)

The release of classified information led to a general court-martial for Nickerson. As a result of the trial, he was reduced in grade and shipped off to serve in the White Sands, New Mexico, zone after a short tour in Panama. The trial in Huntsville received national attention, although the real reason for the court-martial was never mentioned; it was announced that it was about the release of classified information. Nickerson was found guilty, which was a surprise to nobody since the trial was all about the release of classified information—and nothing about the bad decision being made in Washington.

Even though we had been specifically ordered by the Secretary of the Army to desist from doing any work on an artificial satellite, we continued to set up a computer program which would track a satellite after it was placed in orbit.

In two or three instances, the General Accounting Office (GAO) was sent to Marshall Space Flight Center (MSFC) to make sure the orders of the secretary were being carried out. I was always given a few hours notice so that program sheets and card decks could be hidden away.

As it turned out, it was fortunate we continued to perfect the tracking program. It remained as ready to go as the Jupiter-C and was used to successfully track *Explorer 1* in January 1958.

The Thor became the free world's first operational intermediate range ballistic missile when turned over to the United States Strategic Air Command and deployed to England in early 1958. The Jupiter was deployed to Turkey in 1959. Both the Thor and the Jupiter became better known through their use in early space endeavors. The Thor was reconfigured as a launch vehicle to place payloads into space for the air force and NASA. The Jupiter-C was used to place the first U.S. satellite into space in January 1958.

Chapter 12

Challenges and Growth in the Computer Field

In the mid-1950s, computing was struggling to keep up with increasing expectations and the demands of engineers and scientists who were trying to stay ahead in the cold war and to produce the goods which Americans had done without during the war years. The country was between hot wars, Korea and Vietnam, and President Eisenhower was trying to bring peace and harmony to the country. As in the case of most military leaders who have seen the tragedies of conflict and who have had to order men to their deaths, President Eisenhower set about to avoid further conflict after settling the Korean War.

President Eisenhower was concerned about the growth of the federal government, and in particular, the increase in the number of federal employees. The federal government established a limit on the number of civil service employees each agency could hire. This had the effect of placing a freeze on hiring, which became a serious handicap to a new and expanding agency such as the Army Ballistic Missile Agency (ABMA).

To overcome this difficulty the army decided to hire contract employees to fill the gap and to encourage some civil service employees to transfer to contractor groups, thus freeing some civil service slots to be used for more sensitive positions.

Doc Hoelzer planned to expand the Computation Laboratory by hiring additional IBM contract employees, but he ran head-on into a Supreme Court ruling, referred to as the 1956 Consent Decree, which came about as the result of an antitrust suit brought against IBM.

The ruling, which would have wide-ranging effects, required that IBM offer their computers for sale (rather than just for lease) and that they unbundle (price separately) hardware and software. Further, the customer would be allowed to buy cards and tapes in the most economical manner, and IBM would be required to provide maintenance service comparable with what was already being received from the company.

Since its entry in the computer business, IBM had provided computer software and programming services as part of the hardware costs, thus locking the customer closer to IBM.

To accomplish the separation of hardware and software required by the court ruling, IBM formed a new company called the Service Bureau Corporation (SBC). The SBC would be owned by IBM; however, the SBC could have no employees who had a direct connection with IBM and there were to be no advantages in the marketplace accrued to the SBC because of its ownership by IBM.

The original suit had been brought because, in just five years since the visit Doc Hoelzer and I had with N. C. Williams in Birmingham, IBM had become the dominant supplier of computers in the world. Some, particularly the competition, believed that IBM used predatory marketing practices.

Doc had to look for another company to provide him with the programmers to fill the need for the many tasks facing the Computation Laboratory. Herb Grosch (who was running the data center for GE in Phoenix, Arizona) heard of the requirement and came roaring into Huntsville to take over and save ABMA as only Grosch, a well-known and controversial computer pioneer, could do. GE was awarded the initial contract to provide fifty persons to do programming and computer operations.

The existing group in the Computation Laboratory numbered about fifty people and it was hoped that as many as forty would transfer to the GE contract, thus freeing these spaces so that rocket engineers could be hired in their spaces.

Only twenty-five of the civil services people volunteered to go with the GE contract, along with Helmut Sassenfeld, who headed the group in civil service and decided to take the job with GE as manager of the contract group.

As managers of the three branches of the laboratory, Chuck Hubbard, Carl Prince, and I decided to stay with the government—and it would not have been possible to have drawn up a more contentious division of responsibilities even using the plans of politicians when drawing up congressional districts.

A major point to be settled in this early facilities management contract could be defined as a conflict between the concept of facilities management and personal services. Those twenty-five people who stayed in civil service thought that GE would furnish employees as needed, and they would work under the direction of the civil service employees. On the other hand, GE thought they had been given the mission to manage the laboratory and they would decide how best to do it. The concept of intermingling civil service employees with contract employees was then, and still is today, a situation to be carefully monitored. The first years of this arrangement brought many loud discussions, visits with contracting officers, near fist-fights, and threats of resignations.

The situation was greatly exacerbated since Doc Hoelzer faced a dilemma worthy of King Solomon. On the one hand, he had Chuck, Carl, and me; and on the other he had Helmut Sassenfeld, a brilliant young man who had completed his PhD in Germany after the war and had come to the U.S. to study at the Thomas Watson Institute in New York. Doc had been put in contact with Sassenfeld, had brought him to Huntsville, and had helped him to become established in America. He became one of Doc's own.

I was deputy director of the laboratory as well as the contracting officer's representative (COR) on site. The contracting officer was based in Phoenix and on my visits for advice and guidance, his remedy was simple. The contracting officer told me that the basic rule was that GE did the work, and I smoked the big cigar, but it was not working out that way. I was continually urged by my civil service colleagues to take some action, and I was continually ignored by the contractor (GE).

After three years of skirmishes, I finally tried to define the situation and asked for help. The following memorandum to Dr. Hoelzer was executed in 1960.

August 18, 1960

Subject: Operation Mode, General Electric Contractor for Computation Division, MSFC, NASA;

1. During the past three years there have been various opinions by various groups as to the mode of operation by the General Electric group in the Computation Division. These opinions have varied all the way from a complete mission type contract (as in the case of building a piece of hardware) to the complete personal services type, commonly called flesh-peddling. It has been quite clear to the management of this Division that we must come to a stable position somewhere between these two extremes. This memorandum purports to describe the mode which must be followed if the success of the Division within MSFC is to be assured.
2. MSFC is managed by Civil Service employees. Dr. von Braun is the Director of the Center and nine division and three offices report to him on the technical side. These division directories and office chiefs are all civil service employees and form the policy board of the center. It is quite clear no contract group can have representation on this board and no input or output can take place other than through the Division Director for which the contractor works.
3. With regard to the General Electric Contractor within this Division, the following must be done:
    a. The Division (Civil Service employees) must be completely aware of the work being done by the Contractor on a continual basis. There should be no

work performed by the Contractor without full knowledge of the Division.

b. The Division must be able to evaluate work being done by the Contractor. This means the Division must maintain a strong in-house capability and in fact must do part of the work.

c. The Division must act as a buffer between the Contractor groups and the other divisions. This does not imply that day-to-day working groups be prohibited but that a general buffer be established to monitor unreasonable demands in either direction.

d. The Division must report periodically to the user on status of work etc. Under our Center policy no direct funding is made for computer time: therefore, each Division Director must monitor the demands of his people on the Computation Division.

4. To accomplish the things in item 3 the following action has been taken.

   a. The responsibility for scientific digital applications has been given to the digital projects branch. This Branch will organize to perform technical supervision of the contractor and will also maintain an in-house capability.

   b. The responsibility for commercial applications [*in the '50s administrative applications were often referred to as commercial applications*] has been given to the ADPS Branch. This Branch will organize itself to perform the functions stated in 4.a. above.

   c. The Division has been allotted 185 contractor spaces and practically no additional civil service spaces. This Division must use allotted civil service and contractor spaces to best perform the mission assigned. This will demand that the contractor hire people not directly associated with digital computing.

Charles L. Bradshaw
Deputy Director, Computation Division
& Contract Officers Representative

In a day or so, I received a curt, handwritten reply from Doc, which I still have in my files and read when life gets difficult. This may have been the only time in our forty-five years of association where the blood pressure rose between us.

His reply was in the form of a simple organization chart with boxes showing the GE branches (scientific and administrative) reporting to both the local GE manager as well as to the civil service branch chiefs. At the bottom of the organization chart, all in Doc's handwriting and in red, were the words: *"This is it, damn it!!"*

I read his reply and thought, "So much for wordy memoranda, the world ain't that complicated."

The difficulty we were having with integrating support contractors into the division would become evident at a later time. Integration is much easier to do when the mission of a group is known and they are about developing a well-defined product or service. It is much more difficult in a research and development effort.

A few years later, a young man named Ross Perot grasped the idea and formed a very successful company to do facilities management of computer and data related groups.

Electronic Data Systems (EDS), Computer Sciences Corporation (CSC), and University Computing Company (UCC) were the prominent software services firms of the 1960s. The three led the way and saw hundreds of software companies developed which in the late 1960s through the 1970s began to dominate the computer business (over the hardware companies) in income and investment interest.

Electronic Data Systems

H. Ross Perot was born and grew up in Texarkana, Texas, and was a dynamic entrepreneur from his days as a youth. He entered the United States Naval Academy in 1949, graduated with honors in 1953, and spent four years on active duty serving aboard a destroyer and an aircraft carrier.

In 1956, Perot married Margot Birmingham from Greensburg, Pennsylvania. After his service with the navy, they moved to Dallas in 1957, where he became a salesman with IBM. During the early years of their marriage, his wife taught school. The story is told that she loaned Perot one thousand dollars to go into business in 1962.

The company was given the name Electronic Data Systems (EDS), and went further and faster than any computer company in the 1960s and 1970s.

In the beginning, EDS concentrated on doing what was referred to as facilities management. EDS would take over the entire computer operation of a company—including the responsibility for organizing the input and guaranteeing the output from the company's computer systems. For instance, EDS would assume the responsibility for all the data processing required to administer Medicaid for an entire state.

EDS would also take over the data processing systems for an entire company. They were so successful that they were able to get a contract to operate and manage the entire computer operations of the General Motors Corporation (GM).

This worked so well that it became advantageous for GM to purchase EDS. This purchase made Perot a richer man and gave him a seat on the GM board of directors.

Perot did not plan to just sit on the board of GM; he intended to become an active part of management. He pushed for improvement in the products of GM and a plan to extend automation in the plants. He became a burr under the saddle of Roger Smith (GM chairman), leading to a decision by GM to buy Perot's stock for $700 million and to remove him from the board.

Ross Perot was an innovator in many areas. A slogan used by EDS in recruiting personnel was: "Eagles don't flock, they fly one a time"—and Perot was seeking the eagles for employees of his company. The eagles would not follow the crowd, but fly high and look for new opportunities in unexplored areas.

While most companies went to universities to recruit students from the computer science or business departments, EDS went to the registrar of the university and asked for the names of the best students regardless of major. They would take a good student with a major in geology, French, or computer science; send them to a few months of school in Dallas; and turn them into EDS employees, not computer programmers. Perot also established a strict code of conduct for EDS employees, reminiscent of the codes established by Thomas Watson Sr. for IBM, except for the blue suit and the red tie.

Computer Sciences Corporation

Three years before EDS was formed, Fletcher Jones and Roy Nutt left their employers and formed a computer software firm they named Computer Sciences Corporation (CSC). Jones had been a premier marketing whiz for North American Aviation Corporation and Nutt had been a programming genius with IBM. Nutt had developed an assembly program to simplify computer programming, which was widely accepted as the best and had been a member of the team that had developed the FORTRAN programming language for IBM.

CSC chose for its company logo the standard symbol for the atom, showing a group of electrons in orbit around the nucleus. It gave the impression that CSC was formed to support areas of scientific research, and as it turned out their first significant contract was with a scientific installation. In 1961, CSC won a contract to support the NASA Jet Propulsion Laboratory's (JPL) Flight Operations, launching CSC into the space business.

In the next year, CSC won the contract to support the Computation Division of NASA's Marshall Space Flight Center.

In 1964, Bill Hoover, who for a number of years headed the computer activities for JPL-NASA, became chairman and president after the untimely death of Fletcher Jones in a crash of his airplane. Hoover and his associates led CSC to become recognized as the most successful, long-running computer software firm in the world. Some forty-five years after their founding, CSC had grown from two men and a one-hundred-dollar investment to a global giant with more than 78,000 employees in eighty countries. They are still rated among the top 200 U.S. corporations in the Fortune 500.

University Computing Company

Sam Wyly graduated from the University of Michigan in 1957, and worked for IBM and Honeywell for several years prior to establishing the University Computing Company (UCC). Wyly, along with his brother Charles, founded UCC to do computer services in 1963. The Wylys picked that name since it was originally established on the campus of Southern Methodist University, where there was a cooperative effort to manage the computer resources of the university and to market surplus resources. UCC went public in 1965 and made a lot of investors happy when it earned them a 1,000 percent return on their investments.

The Wylys, like Perot, lived in Dallas, and the alleged race between UCC and EDS made good press for the trade magazines. UCC was able to get several large government contracts and several contracts with the oil companies in Houston, and was off and running.

As part of its growth and impact, the computer industry also began to develop as a profession. The first professional organization to be devoted to the development of computers was the Eastern Association for Computing Machinery (EACM). This group was formed in 1947 shortly after the ENIAC was unveiled and other computer development was under way in Boston and Philadelphia. The first meeting of the EACM was held at Columbia University on September 15, 1947, with seventy-eight attendees. The name was changed within about a year to the Association of Computing Machinery (ACM). The second meeting was held at the Aberdeen Proving Grounds, and there were three hundred attendees.

Other professional societies formed over the years. The Data Processing Management Association (DPMA) was formed from the older National Machine Accountants Association (NMAA) which had been an association of persons using tabulating equipment and punched-card machines for data processing. The DPMA has concentrated its interest on administrative applications, as opposed to the ACM which has focused primarily on scientific, engineering, and more novel applications.

The ACM published the first edition of the *Journal of the ACM* in January 1954. This was the first scholarly journal devoted to the development of computers and their application to problems in science and engineering. The ACM also published several other periodicals which expedited the distribution of developments taking place in computer hardware, software, and the ever-growing area of computer applications.

In addition to the ACM publications, periodicals appeared which were supported by advertisements or subscription fees. The first trade magazine devoted to computing was entitled *Datamation*. The first edition of *Datamation* was published in 1955, and was well received throughout its forty years of existence. A computer-related newspaper called *Computerworld* came along in 1964.

In the mid-1950s, administrative applications began to outgrow the capability of the EAM equipment. Efforts were undertaken to program some of these applications for the scientific and engineering computers available in their organizations.

There was one very cumbersome problem for those doing administrative applications on the IBM 704. The 704 was a word machine, meaning that an entire word of thirty-six bits was operated on as a unit. As administrative applications were moving from EAM to the computer, this became a serious shortcoming. Administrative applications needed to manipulate characters (the word "byte" had not yet been invented). IBM was quick to remedy this problem by reconfiguring the 704 to meet this need, renaming it the IBM 705. Doc Hoelzer's laboratory installed two 705s and off-loaded all the administrative applications to the newly formed unit of the laboratory which was named the Administrative Applications Division. Bill Fortenberry was hired to administer this new and rapidly growing division. He became one of Doc's favored boys.

The installed IBM 704s at ABMA were fully loaded. As research and engineering took place to finish the Jupiter ahead of the air force's Thor, engineers scrambled for more computing time. Batched jobs were shipped by air along with a programmer to computer sites in Atlanta, Phoenix, and Washington, D.C., to be processed on weekends as computer time became available.

Every afternoon at four o'clock, there was a mad scramble at the dispatch desk in the computer room. There was no operating system on the 704, and even more inefficient, there was no overlap of operations. If cards were being read or output being printed, all other components were idle. If it took five minutes to read the input, then forty-five minutes to do the computing, and then ten minutes to do the output, the job would take a full hour.

The computation laboratory learned of small, desk-size computers and decided to procure some of these and place them in selected research areas. The LGP-30, built by the Librascope Division of the General Precision Company, was one of the earliest desk-size computers. It became very popular, and approximately five hundred units were sold—a large number for 1956–57.

In designing the LGP-30, Librascope had kept the number of vacuum tubes to a minimum for the sake of overall reliability and used solid-state diode logic where possible. At the time it was considered very reliable. Some researchers started a problem on Friday afternoon, checked it on Saturday morning and perhaps on Sunday afternoon, and would come to work on Monday morning expecting to find the results from a weekend of computing.

The LGP-30 had sixteen commands in a single address system and had storage capacity for 2,048 thirty-two-bit words on a magnetic drum. The capacity of the drum was later expanded to 4,096 words. Input-output was by punched paper tape. An ALGOL-like high-level language called ACT-111 was developed for the computer. The entire system was the size of a large desk. In 1957–58, there were more than twenty LGP-30s in use at ABMA.

Another significant small computer was the Bendix G-15. This computer was designed by Harry Huskey, a prominent early computer scientist, and was said to be influenced by the design ideas of Alan Turing. The Englishman Turing is credited in many quarters as being the person who developed the concept of the modern computer in 1936 with his concept which is now called the

Turing machine. The Bendix G-15 had moderate success and led Bendix to design and market a larger computer called the G-20.

There were several other desk-size computers introduced. The Underwood Typewriter Company marketed one they called the Elecom-100. Royal McBee Corporation introduced a computer which did not have great success. The Autonetics Division of North American Aviation sold several small computers they had named the Recomp-1 and the Recomp-11. A little later, the IBM 1620 had great success and ushered many colleges into the computer age.

In the mid- to late 1950s, there were thirty to forty desk-size computers in use at ABMA. These computers were purchased (or leased), maintained, managed, and programmed by the Computation Laboratory. Where possible, general purpose programs were written which could be passed around from one user to the other. Examples of these types of programs are those used to find the roots of polynomials and programs to solve simple differential equations.

The success of the desk-size computers led to the concept of laboratory computers. There were nine laboratories (major organizational units) in the technical organization headed by Dr. von Braun, and five of these laboratories made requests for a computer. The computers which were of the dimension being considered for laboratory use were the GE-225, the Electrodata (Burroughs) 205, the CDC 1604, and the IBM 650. These were installed in the five laboratories and were provided by, programmed, and operated by the Computation Laboratory.

The concept of central control of computers continued to be an area of stress and heated discussion. In one of Dr. von Braun's staff meetings, a laboratory director made a strong pitch for having his own computer without control of the Computation Laboratory. He seemed to be making some progress—until Doc Hoelzer came to his feet and said that if the Aeroballistics Laboratory was given its own computer, then he wanted a wind tunnel for his laboratory. As the discussion became more heated, Doc jumped to his feet again and stated loudly that he was quitting if another laboratory was given a computer. This led Dr. von Braun to state, "Helmut, sit down and be quiet. You can't quit, you are too old to find another job." Doc gave a weak smile and sat down.

Doc had a temper which came through at times when he felt things were not moving as quickly as they should. In one instance, I was outside his office talking with Fran Mitchem, his secretary. Doc was on the phone, and the volume of his voice was gradually increasing. He was upset about something, and the conversation became louder and more serious. The loud talk on the phone ended when Fran and I heard a crash which sounded like a telephone hitting the wall.

Doc charged out the door and, seeing me, gave me a direct order. He exclaimed, "Brad, go in there and get on the phone with that idiot and tell him that I have gone home and I never want to talk with him again."

I did not know what idiot was on the phone, but I crawled across the room and found the phone. An engineer from the facilities office was still on the line. I mumbled something about Doc having had a bad day and things would be okay by tomorrow.

The engineer laughed and assured me he was not upset. Doc was one of his best friends and he knew Doc would be in his office the first thing the next morning. Doc would apologize to him, bring him a candy, and they would have a good talk, a good laugh, and a cup of coffee.

One day Doc was running late for a meeting with Dr. von Braun. Doc lived on Monte Sano Mountain, which bounded Huntsville on the east side, and he had to go through downtown Huntsville to get to the arsenal. He came off the mountain at a good speed and was continuing this speed through the downtown area. Doc was pulled over by a policeman for speeding, and the officer began the process of giving him a ticket.

Doc explained to the policeman, in a very polite manner, that he was on his way to the arsenal for a very important meeting with Dr. von Braun and he did not have time for the officer to write him a ticket. But, Doc told the officer, if he would be at this same location the next day at the very same time, Doc would come through, driving at the very same speed as today, and then he would have time to let the officer give him a ticket. With that, Doc drove off for the very important meeting.

The officer had never seen anything like this, was stunned, and had no procedure in the rule book to handle this situation. There were no repercussions, and I never asked Doc if he kept his end of the bargain on the next day.

During the early days of ABMA, Huntsville was a lot like Oak Ridge, in that you had to belong to the Elks Club or the American Legion if you wanted an alcoholic beverage with your steak. These two clubs saw a record number of new members, some from as far away as California, since marketers and promoters were moving to Huntsville like shellcrackers to a willow fly hatch.

There were a lot of after-hours marketing sessions held at the American Legion, and in some cases, the drinking went into the late evenings. In a few instances, the drinks won, and a person succumbed to alcoholism. This happened to a manager in Doc's organization who went through various stages including the delirium tremens. His wife reported this to a couple of friends, and asked that they stay away because her husband was going through a stage of fear and was armed to protect himself from imagined enemies.

Everyone kept this from Doc for a couple of days for fear he would fire the person, but the word got to him. Within a few minutes, he was in my office and said, "Brad, get your jacket and come with me. We are going to get those guns before someone gets hurt." As I followed Doc out to his car, remembering the dire warnings from the wife about the armed camp we were entering, I wondered why I wasn't a little frightened.

When we arrived at the home, the front door was unlocked because Doc had his secretary call the wife to warn her that we were on our way. We went directly into the house and our friend was standing, stunned at the sight of Doc, in the middle of the living room. He was certain he was about to be fired in person by Doc, who went directly to the point without a howdy do, and said we had come to get the guns.

As the man mumbled something about not knowing about any guns, his wife glanced toward the adjacent bedroom. I proceeded to get down on the floor and pull out two long guns and a pistol from under the bed. They were loaded.

As I carried the guns to the car, Doc told our friend he was missed at work and we needed him there—that all his good friends needed him there. Doc said he would put him on sick leave until the following Monday, and that he was to come in and report to Doc on that day.

The next day, the wife called Doc to say her husband had checked into a hospital, and he promised to be back at work as soon as he was released. Doc put him on sick leave with pay, and saw the man return as promised. He continued to help him through some difficult months to return to good health and to remain a valued employee and good friend.

CHAPTER 13

THE RACE IS ON: FROM JUPITER-C TO EXPLORER 1

The Vanguard rocket program of the navy was becoming a spectacular failure. People gathered at the watering holes along Cocoa Beach for scheduled launches of the Vanguard rocket just to see the fireworks when the explosion took place. The nation had asked too much from a group of inexperienced rocket scientists, and the project went down as one of the worst decisions of the Eisenhower administration.

As the U.S. satellite program failed, Russia was making good progress because they were using a modified military rocket in which the basic engineering had already been tested. They had a much simpler task of modifying a working missile to give it the added thrust necessary to place a satellite into orbit.

In early October 1957, I was on the couch taking my usual Sunday afternoon nap, when there was a panicked knock on the door. In came Chuck Lundquist, full of emotion and great disappointment, blurting out that the Russians had done it—they had placed the world's first artificial satellite into orbit. He said we should go to the arsenal and listen to a tracking station to see if we could get enough data to compute the path of the Russian satellite with the unauthorized computer programs we had written.

We did. It was an eerie feeling to listen every ninety minutes to the beep-beep of a satellite placed into orbit by an enemy who had threatened to annihilate the United States.

President Eisenhower announced later that day that the Russian satellite represented a neat scientific achievement but posed no threat to our security. He was, of course, correct. Nevertheless, the U.S. immediately became a second-rate power in the eyes of the world. This inferior status would not be overcome until 1969, twelve years later, when Neil Armstrong stepped out on the moon.

There was suddenly a great urgency to place a satellite in orbit. The secretary of the army once again approached the administration to listen to the proposal from the team in Huntsville; General Medaris and Dr. von Braun were called to Washington. General Medaris made the proposal to use the Jupiter-C rocket to launch a twenty-eight-pound satellite, stating it could be done in ninety days. The army was given the order to proceed in mid-November; long work hours and great excitement followed.

The Jupiter-C, which had been used to solve the reentering warhead problem on the Jupiter, consisted of the Redstone with elongated tanks for more fuel and longer engine burn-time, plus a configuration of nine small solid-propellant rockets as a second and third stage. After the elongated Redstone had placed the second stage into the correct position, the Redstone was jettisoned and eight of the small solid-propellant rockets ignited and began to spin for stability.

The proposed satellite was to be added in the middle of the spinning rockets with an additional solid-propellant rocket to hurl it into orbit at a speed of 17,500 miles per hour (mph). A variation in velocity above 17,500 mph would determine the type of elliptic orbit which would be attained. A small amount less than 17,500 mph might cause the satellite to return to earth. A speed near 25,000 mph, called escape velocity, might cause the satellite to escape the earth and become a satellite of the sun, never to be heard from again. A specific window had to be hit.

The engineers at MSFC quickly pulled a Jupiter-C out of storage and went into a cooperative effort with the Jet Propulsion

Laboratory (JPL) in Pasadena, California. JPL had built the upper stage of the Jupiter-C and had other capabilities helpful in designing and tracking the satellite once in orbit.

The job of building the satellite and its communication system was assigned to JPL. They worked with James van Allen of the University of Iowa in designing the onboard experiment to measure the intensity of the radiation belt surrounding the earth.

During the period when we were preparing the computer programs to track the satellite, Chuck Lundquist and I had several meetings with Prof. Leland Cunningham in Berkeley, California. Professor Cunningham was an outstanding astronomer as well as a pioneer in the use of computers.

We were young, enthusiastic, and in Professor Cunningham's view, somewhat naive. During one of our meetings, he looked us straight in the eyes and said, "Now, I have a lot of respect for the enthusiasm of you young men, and I want to help you all I can. You may get a satellite into orbit, but you will never get a man into space." There was a general perception among many astronomers that the earth was surrounded by strong radiation belts (and perhaps other barriers) which would not allow man to survive the trip into space.

The excitement and dedication of the von Braun team had never been greater, and it was as if all America was looking to Huntsville to get the United States back in the space race. In addition to the engineering and checkout of the launch vehicle, there were other problems, since the electronics and telemetry devices were not as sophisticated as they were to become a few years later, and there was no onboard digital computer to help with decisions.

A big problem was to determine when the upper stages were in the right position to fire the rocket to attain orbit. There was the possibility to read some telemetry in the blockhouse and make an educated guess as to the right time to fire the second stage. Dr. Ernst Stuhlinger was given this difficult and all-important task.

A task given to the computing group was to determine the orbit (path) of the satellite. Chuck Hubbard had been placed in charge of this effort and was assisted by Jud Hart, Jim Golman, and Mildred Letherwood. Fortunately, we had been disobeying orders from Washington and had the computing program ready to go. Hubbard did some last-minute checks and corrections after conferring with Professor Cunningham. The initial conditions for the orbit program would come from Cape Canaveral and would be corrected based on data from tracking sites. The computation group was ready and anxious for the launch.

It was January 31, 1958. The Jupiter-C was on the pad at Cape Canaveral; the *Explorer 1* satellite sat atop the second stage and was "talking back." The Missile Firing Laboratory was ready to go; the computing group was in Huntsville with computers standing by for data; phone lines were open to the Cape and to the tracking sites. The whole free world waited to see if the United States could put a satellite in orbit and get back in the space race.

In the early evening, the countdown proceeded to zero and there was ignition of the main stage. It was a flawless liftoff. The guidance system led the missile perfectly down the Atlantic Missile Range while Dr. Stuhlinger studied every telemetry device available in the blockhouse. Cut-off of the main-stage engine took place right on schedule, and the stage and the fuel tanks were jettisoned. A coast period of a few seconds was scheduled for the second stage, and at precisely the right time a button pushed by Dr. Stuhlinger ignited the upper stages and hopefully placed the satellite in orbit.

We could not be certain the satellite was in orbit until *Explorer 1* had made a complete revolution of the earth and its signal was received by the Goldstone Tracking Station in the desert east of Los Angeles. The tracking site at Antigua at the end of the Atlantic Missile Range reported it had picked up a signal that indicated the upper stages had fired correctly. There were two or three tracking sites around the world which were standing by in case they were able to pick up the signal from *Explorer 1*. All we could do was to wait and hope.

Dr. von Braun; the head of JPL, Dr. William H. Pickering; and Professor van Allen had been ordered to Washington to participate in a press conference after the flight and were standing by in the Pentagon. General Medaris was in the blockhouse at the Cape with Dr. Debus. Lundquist, Hubbard, and their computer crews were in the Computation Laboratory in Huntsville awaiting data so they could establish the path of the orbit. Hundreds of other employees of JPL and ABMA were standing by to help and, they hoped, to celebrate.

The first turn around the earth for *Explorer 1* was computed to take approximately one hour and forty-five minutes. Dr. Pickering had an open phone line to the tracking station at Goldstone. At the expected time, there was no word from the tracking site; Washington officials were getting frustrated and they began to look to Dr. von Braun for answers.

After five minutes, Dr. Pickering spoke rather loudly and anxiously into the open phone line and asked his employees at JPL if they had received a signal from *Explorer 1*. There had been no signal. There was concern that *Explorer 1* had not made it into a proper orbit or that its transmitter was not working.

Shortly thereafter, word came from Goldstone—*Explorer 1* had been acquired. Then we all heard General Medaris speak from the blockhouse at the Cape, "Goldstone has acquired. Gentlemen, we are in orbit." The celebration, the news conferences, and the computing began.

The celebration began in the blockhouse at the Cape, then all the sites at ABMA in Huntsville, the JPL sites in Pasadena, then downtown Huntsville, then all over the free world. People left their homes and flowed into the streets. In Huntsville, bands were quickly formed and a parade was started. The mayor was on every platform making speeches. The town which seven years earlier was known as "the watercress capital of the world" would now become "Rocket City, USA."

The satellite tracking programs for earth satellites written for the 704 worked extremely well. These programs had been able to plot and predict the path of *Explorer 1* with the required precision. Chuck Lundquist had his staff prepare a display for Medaris which showed the location of the satellite above the earth on any day at any time. The equations, prepared by Lundquist and scaled and programmed by Chuck Hubbard and his group, worked to perfection. The magnetic core memory of the 704 had performed without failure.

There was an occasional problem with a mechanical part of the computer, but compared with previous computers the performance was a great success. There was not much talk about mean time between failures (MTBF), and the phrase which had existed since the earliest days of digital computers was disappearing from everyday conversations.

The IBM 704 and 705 had one major design flaw which had been recognized from the beginning; the computers could not use their various components in parallel. If the card reader was reading cards, no other part of the computer could be used, and the CPU and the other components sat idle while large decks of cards were entered into the computer. The same was true of the relatively slow line printer; thus a very expensive CPU sat idle while routine and slow components such as the card reader were in use. This problem had been recognized by IBM, as well as by most of its users, and a program was initiated to solve the problem. The result was a new but compatible computer which IBM named the 709 (using IBM arithmetic 704 + 705 = 709). The 709 was both a word and character machine, and could replace both the 704 and the 705, in addition to solving the parallel operation problem.

Satellite Equations of Motion

1956-57

$$\ddot{x} = -\frac{kx}{R^3} - \frac{3kjx}{R^5}\left(1 - 5\frac{z^2}{R^2}\right) - \frac{1}{m} f(v,R)(\dot{x} + \omega y)$$

$$\ddot{y} = -\frac{ky}{R^3} - \frac{3kjy}{R^5}\left(1 - 5\frac{z^2}{R^2}\right) - \frac{1}{m} f(v,R)(\dot{y} - \omega x)$$

$$\ddot{z} = -\frac{kz}{R^3} - \frac{3kjz}{R^5}\left(3 - 5\frac{z^2}{R^2}\right) - \frac{1}{m} f(v,R)\,\dot{z}$$

$$R = \sqrt{x^2 + y^2 + z^2}$$

$$f(v,R) = \frac{1}{2} A C_D v \rho(R)$$

$$v = \sqrt{(\dot{x} + \omega y)^2 + (\dot{y} - \omega x)^2 + \dot{z}^2}$$

$$k = 5.16671 \times 10^{12}\ \text{km}^3/\text{hr}^2$$

$$j = 2.22189 \times 10^{4}\ \text{km}^2$$

$$\omega = .262516143\ \text{radians/hr}$$

m = mass of satellite

A = area of satellite cross section

C_D = drag coefficient, normally = 2

$\rho(R)$ = density function

Above and on pages 204–205: The three second order differential equations, constants, and scaling considerations which defined the path of Explorer I. *Programmed for the IBM 704 computer.*

Smithsonian Density Function

$$\log_{10} \rho(R) = 5.588 - 4.79 \log_{10}(9.0167 H - 603.318)$$

$$H = \frac{6356.766\, G}{6356.766 + G} \quad \text{(Geopotential Altitude) km}$$

$$G = R - \left[\frac{B^2 R^2}{\frac{B^2}{A^2}(x^2 + y^2) + z^2}\right]^{1/2} \quad \text{(Altitude) km}$$

$$\left.\begin{aligned} B &= 6356.5838 \text{ km} \\ A &= 6378.2064. \text{ km} \end{aligned}\right\} \text{Clarke Ellipsoid}$$

Units of $\rho(R)$ is Kg/m^3

Calculation of $\rho \cdot 2^{26}$ (a machine number < 1)

$$\frac{\log_2 \rho}{60} = .30938 - .079833 \log_2(9.0167 H - 603.318)$$

(a negative machine number)

Define $Q = \frac{\log_2 \rho}{60} + \frac{26}{60}$

$$Q = -.56032 - .638664\, \overline{\log_2}(.90167\, \overline{H} - .0073647)$$

where $\overline{\log_2} = (\log_2) \cdot 2^{-3}$

$\overline{H} = H \cdot 2^{-13}$

$$\rho \cdot 2^{26} = 2^{60Q} = 2^{-n} \cdot 2^{-\alpha}$$

where n is a positive integer
α is a positive fraction

Procedure is to calculate $2^{-\alpha}$ and right shift the result n places.

SCALING CONSIDERATIONS

x, y, z, R scaled by 2^{-14}

$\dot{x}, \dot{y}, \dot{z}, \dot{R}$ scaled by 2^{-16}

$\ddot{x}, \ddot{y}, \ddot{z}$ scaled by 2^{-18}

Thus if Δt in the integration procedure is scaled by 2^{+2}, the integration program is independant of scaling considerations.

Scaling of Kepler term

Scale k by 2^{-46}

Then $-\dfrac{(k \cdot 2^{-46})(x \cdot 2^{-14})}{(R \cdot 2^{-14})^3}$ has scale 2^{-18}

as required for $\ddot{x}$

Define $\bar{k} = 2^{-46} k$; $\bar{x} = 2^{-14} x$; $\bar{R} = 2^{-14} R$

The new 709 computers were program compatible with the 704 and the 705 so there was no need to change any of the existing computer programs. However, the one significant design objective of the 709, allowing for an overlap in the use of the different components, would require a major program development. For the first time in the IBM line of computers, the input devices, the arithmetic units, and the output devices could be used simultaneously.

This brought about the need for a master program which would schedule the three components of the computer for maximum efficiency. This master program—known as the supervisor program or the executive program, and later known as the operating system—would schedule the jobs to be done and allow for the maximum use of the input system, output system, and compute system.

The Computation Laboratory of ABMA decided to develop an operating system, and the job was given to Bill Sullivan of the GE systems staff. IBM provided Danny Kileen, an excellent systems analyst, to assist with the project. Danny and Bill, assisted by two other programmers, developed one of the earliest operating systems. It was named Supervisory Program Over Other Kinds (SPOOK) and was used with good success for several years at the ABMA Computation Laboratory. It became known as one of the pioneering efforts in the development of computer operating systems.

While sitting in a meeting with Doc Hoelzer, I noticed a very small nodule in my neck just above my collarbone. I assumed it would go away but it didn't. In a few months it grew to about the size of a golf ball.

In November 1958, I visited with a surgeon in Knoxville, Tennessee, who recognized there might be a problem. Exploratory surgery led to the conclusion that there was a serious problem; the nodule was a thyroid cancer, which led to major surgery in January. The surgery was performed in Knoxville (approximately two hundred miles from Huntsville), and the doctor recommended that after five days of recovery in the hospital, I should go home by commercial aircraft.

Bernice and I arrived back in Huntsville on Friday afternoon, and as the plane taxied up to the gate at the airport, I saw Doc and a group from the Computation Laboratory waiting to meet us. In those days, the plane simply taxied up to the airport building, and the passengers walked down the ramp in the back of the plane. I got hugs from all the ladies and a good handshake from Doc, Chuck, Carl, Bill, and some others.

When the greetings were over, Doc took me aside, gave me some papers to study over the weekend, and said he hoped I would be at work on Monday morning. He explained that some of the projects I was responsible for were getting out of hand and far behind. I could not imagine what these projects were, but I thought a lot about this over the weekend and worried that my job might be in jeopardy. This would be a bad time to lose my job, so I spent a lot of time studying the papers Doc had given me and made sure I would be ready to go when I returned to work.

On Monday morning, Doc laid out some more projects having to do with the computers we were installing and wanted to know how soon I could travel to Cape Canaveral and look into a computer organizational problem which was developing.

It occurred to me that Doc was taking his job much more seriously than I had seen him do in the past. I wondered if he was in jeopardy of losing his job, and this caused me more concern.

After some three weeks of hard work and worry, it occurred to me that Doc had a secret plan. He knew the worst problem with cancer was the worry over whether or not the cancer had metastasized and was hiding away in some vital organ ready to take you down at any time. It occurred to me that I had not given much thought to the cancer in the three weeks I had been back at work. My main worry had been whether or not my job was secure—and that worry was a lot less serious than worrying about dying and leaving three small children with no father. This incident was just one of the many times I saw Doc give considerable thought and planning to promote the welfare of people with whom he worked, particularly those associated with the Computation Laboratory.

A few weeks before my illness, Dr. Oswald Lange, the deputy to Doc Hoelzer, had been promoted to a staff office. Doc needed to appoint a new deputy and the choice was logically either Chuck, Carl, or me. Doc never wanted to get in the position of having to choose among the three of us. My illness gave Doc a chance to proceed with what he wanted to do: promote me to the position of his deputy.

Doc called Chuck and Carl into his office and explained that he needed to appoint a deputy from among the three division chiefs. Doc explained that he had made his choice and that he wanted to appoint me to the position because, with my serious illness, it might turn out that I wouldn't make it. Doc continued by pointing out that I had a family with three small children; a promotion would benefit my family in the event the very worst happened with my illness.

Chuck responded that he understood and thought it was the wise thing to do.

Carl, in a rather typical fashion, leaned way back in his chair, took a deep drag on his cigarette, and responded by saying, "Doc, all of what you say is well and good and I agree with you 100 percent, but what if the son-of-a-bitch don't die?"

Neither Doc nor Chuck could come up with an adequate response.

Chuck and I later laughed about this incident many times. He suggested I not go bird hunting with Carl and to wear an extra life preserver while fishing.

A couple of months later six of us went on our annual long-weekend fishing trip to Center Hill Lake in Middle Tennessee. After a day of fishing and fried fish for supper there was the customary poker game at which a bottle of booze always appeared. After a few hands of poker, I slipped off to a bunk for some rest.

In a little while, Carl missed me and yelled, "Where is Brad?" Someone said I had lain down for a while to rest. Carl got out of his seat, stuck his head into the bunkroom and yelled, "Bradshaw, get your butt out here and enjoy this poker game while you can. You ain't got much longer to live anyway so you better play some poker while you are still alive!"

This was too much for Chuck. He braced Carl toe to toe and chin to chin. He said, "Carl, if you don't sit down there and shut up, I am going to take you outside and whip your ass."

Carl was stunned.

The scars on Chuck's face were snow white; the curse of a red-headed guy is a few skin cancers. No one had ever seen him so serious.

Carl said, "Chuck, I believe you mean it." Chuck assured him he did. Carl sat down, said nothing else, and continued to win all the money at the poker game as he usually did, another skill learned well in the navy.

Chuck likes to tell of the time he and Doc were having a serious difference on the approach to a project. Both were firmly entrenched in their approach and unable to persuade the other to change his opinion. Doc would never order Chuck to do it a certain way but would try to convince him to agree with his approach. Chuck also had the ability to stick to his convictions even under strong pressure.

Finally, Doc was overcome with frustration. He threw his hands up in disgust and exclaimed, "Chuck, you are hopeless, and I am through arguing with you. Good-bye!"

Chuck watched from his second-story window as Doc got into his Karmann Ghia, went out the gate, and then stopped. Doc turned his car back into the parking lot and was soon at Chuck's office door with a big smile on his face. He said, "Chuck, today is Friday and I want to tell you to have a good weekend. We will continue our discussion on Monday, yah?" Doc did not want Chuck to spoil his weekend with worry over the disagreement they couldn't settle.

Chapter 14
The Solid-state Revolution

On a Friday in late 1958, the local IBM representative came to my office and asked that a very important meeting be scheduled for Monday morning. Dr. Hoelzer was in Germany on vacation, and as his deputy, I was expected to act in his stead on this very important matter. The IBM representative had indicated IBM would be sending its top official from Atlanta and another from New York. It sounded urgent and gave me something to worry about over the weekend.

At precisely nine o'clock on Monday morning, four representatives from IBM came into my office dressed in typical IBM fashion—blue suits, conservative red ties, and white shirts. They sat in the order they appeared on their organization chart. On my right was the local salesman, followed by the local branch manager, the regional manager from Atlanta, and finally the man from the corporate office in New York.

You could identify where each stood on the organization chart by the quality of his suit. The salesman wore a suit from J. C. Penney; the branch manager wore a Hart Schaffner and Marx from a local department store; the regional manager wore a Brooks Brothers suit; and the vice president from New York wore a suit of a quality with which I was not familiar. I wore my best suit, which came from Belk-Hudson, a local department store.

The local salesman began the meeting. He introduced the local branch manager, who introduced the regional manager and the corporate executive, and then told me that I had been chosen as one of only three people in the world to receive an early announcement of a major event. The president of IBM had picked Gen. Bennie Shriver of the air force, Adm. Hyman Rickover of the navy, and me as a representative of the army as the three persons important enough to receive this earth-shaking announcement.

Some might say IBM used this introduction as a way to put me in fast company and to put me on the spot. However, I was more interested in how Doc would handle this and how many he would order, than I was about being grouped with the general and the admiral.

The regional manger went to my blackboard, pulled out something that looked a lot like a diaper, and very carefully cleaned the board and the chalk tray. I thought it peculiar but realized he was preparing it for the vice president from New York who wore the expensive dark blue suit.

When the regional manager completed the cleanup of the blackboard, he handed the chalk to the vice president, who, without one word, went to the blackboard and wrote in large print: "709 TX."

The vice president looked at me and asked if I knew what that meant. I replied that I knew about the 709 since we had two of them already but I had no idea about the *T* and the *X*.

He responded by telling me that the *T* stood for *transistor* and the *X* stood for *experimental*. He then proceeded to tell me some of the advantages of the transistor over the vacuum tube. The transistor was a solid-state device requiring perhaps as little as a hundredth of the electric power of the vacuum tube. He continued by explaining that the vacuum tube worked by boiling electrons off a red-hot wire mounted in a carefully blown and evacuated glass case and was a failure waiting to happen, while the transistor was solid-state and could theoretically last forever.

The IBM vice president then came to the most important advantage of the transistor over the vacuum tube. With a few thousand vacuum tubes in a large computer, the probability of one failing was high enough to discourage building larger computers which would have many more vacuum tubes. The chance for failure would increase geometrically. With the transistor, the mean time to failure in computer components could be lengthened by a factor of one thousand or more. This meant that the sky was the limit and much larger computers were in the offing.

The vice president asked me what I thought.

I replied that the *T* sounded wonderful but I didn't like the *X* too well. I had said just what the vice president had hoped I would say.

He borrowed the diaper from the regional manager, went back to the blackboard, and with an aggressive swipe, he removed the *X* and told me that at that instant (9:15 AM CST) IBM was removing the *X* and announcing the 709 *T* and taking orders. *How many did I want?*

This was the moment of truth. Well, not too truthful, since in those days you simply placed a letter of intent to purchase, which was not binding and could be changed at a later date. I decided to place a letter of intent for three. I was certain the air force would order more but I reasoned they would need that many to catch us in the race between our Jupiter missile and their Thor. My order pleased the IBM representatives. Even though these letters of intent were not binding, IBM announced the next day in the *New York Times* with their public product announcement that they had received orders for a large number.

The next day, I realized I had seen a major breakthrough in computing. It was much bigger than the news of the square-root Friden. The unreliable, heat-producing, electricity-hog vacuum tube was being replaced, and as it turned out, the replacement of the tube by the transistor opened a great frontier in computing. Many more companies moved into the competition and great innovations began in computing, input, output, and program preparation.

I recalled the day in the late 1950s when I heard Prof. John Vincent Atanasoff of Iowa State University at a meeting of the Association of Computer Machinery. Professor Atanasoff told of his thoughts during the mid-1930s concerning the need for an automatic computer which did arithmetic using digits instead of calculating with analog computers. Professor Atanasoff brought together in his mind what a computer using digits and doing addition and subtraction would need to look like.

He told of his frustrations during the winter of 1938 when his mind was cloudy and he couldn't seem to make progress. One evening he went to his office to try to force some progress but nothing happened. He reasoned that his mind needed, as he said, to be jogged. He reasoned that a good stiff drink of whiskey would work to implement the jog, but he was in the middle of Iowa, which was a dry state with no available liquor. Professor Atanasoff decided to make the long drive to the Illinois border where he could buy an alcoholic beverage.

He remembered driving down this lonely road during the heart of the midwestern winter and the thoughts which were running through his mind. As soon as he crossed the Mississippi River into Illinois, he saw a flashing cocktail sign and pulled in at the first roadhouse. He ordered one drink and felt a lot better. A second drink gave his mind a good jog, and the ideas came to him like firemen sliding down a greased pole.

The first idea was that the computer would have to be electronic since a mechanical device could never give the speed which would be needed. The only electronic device which was available at that time and which could give a fast flip-flop was the vacuum tube.

This led to the next idea that the electronic computer would have to be binary, and the computer would perform binary arithmetic. With these ideas floating around in his well-jogged mind, Professor Atanasoff carefully drove back to Iowa State.

We can only assume Professor Atanasoff made the trip home without incident or arrest. He continued to draw together his plans for a digital computer, and in 1939, Iowa State gave him a $650 grant to start his challenging project. Some of the grant funds went to hire a brilliant electrical engineering student by the name of Clifford Berry. The two continued to work on their computer from 1939 through 1941 until the war years intervened.

It had been twenty years since Professor Atanasoff had his two cocktails lead him to the vacuum tube as the electronic device

Dr. Hoelzer, Eberhard Rees, and Dr. von Braun with the first installed IBM 7090, 1960.

Dr. Helmut Hoelzer with model of 7090 at static test stand for Saturn S-1, *1960.*

which would allow him to build his computer. In 1958 we were ready to give up on the vacuum tube and move into the second generation of computers.

The IBM 709 T was named the 7090 before it was delivered. The first standard 7090 was delivered to ABMA in June 1960. Modified versions of the 7090 were being incorporated into the Ballistic Missile Early Warning Systems at the same time. Within a few months after installation, the systems were modified in the field, adding more memory and speed, and renamed the 7094. The IBM 7094 became the workhorse computer of the space program during the 1960s.

In twenty years, computing had moved from the experimental laboratory to the mainstream of American progress. Computers were creeping into areas of accounting, management, and the arts, as well as in every aspect of research. Computers had become essential to progress in the nuclear and rocket sciences.

The electronic stored-program digital computer has from the beginning had four essential components. A computer needs to communicate with the user; this capability is called the input-output system. The computer must be able to store and retain data and programs; this is called memory.

The memory is generally made up of two parts: main memory and auxiliary memory. The main memory is where data and programs must be before execution can take place. The main memory is usually volatile, meaning the data and programs stored there are lost when the computer is powered down. The auxiliary memory is for storage of data and programs waiting to be used by the computer and for permanent storage when the computer is powered down.

The computer must have a unit to do the calculations, generally called the arithmetic unit. There must be a traffic cop to make sure the various operations take place on time and in good order; this unit is called the control unit.

The principal characteristics of the input-output system are that it be accurate and fast. The first input-output systems utilized devices already existing and available off the shelf; they were not fast but were more or less reliable. The teletypewriter had been in use for many years with Western Union and was readily available; it was used for input on many early computers. For output, the computer punched a paper tape which would then be taken to the slow teletypewriter for printing.

As computers became available from UNIVAC and IBM, the input-output was by punched cards, and devices to process cards were immediately available from both UNIVAC and IBM. In the late 1950s and the 1960s, this was the principal mode of input-output. The printers became much faster; there was even talk of page printers which would print a page at a time.

The basic requirements for memory have always been that it be reliable and fast. The words "random" and "sequential" have generally been applied to types of computer memory. Sequential memory refers to the manner of storing such as on a magnetic tape where the activity is started at a certain point and then the data is stored sequentially. Sequential memory means that the computer has to search a sequence until it finds the desired data.

Random memory means that a piece of data is stored in a specific location with an address, which will allow the computer to get a specific piece of data by referring to a specific location and retrieving the data directly and without a search.

In the 1950s, the Williams tube and later, the core memory, were examples of random memory, while magnetic tapes, rotating drums, and magnetic disk platters were examples of sequential storage.

Started in 1944, the ENIAC used the vacuum tube so that a ring of ten flip-flops made a ten-decimal digit (0–9). A ten-decimal number required one hundred flip-flops. The ENIAC was built with a total memory of only twenty ten-digit numbers, and these required two thousand vacuum tubes. The arithmetic operations on the ENIAC consisted of a multiplier, a divider, and a square rooter.

A master programmer initiated and controlled the sequence of operations, requiring more vacuum tubes. When completed, the ENIAC had eighteen thousand vacuum tubes. The ENIAC required so much electricity that the citizens of west Philadelphia claimed their lights dimmed when the computer was turned on. Vacuum tubes were not reliable, and if only one failed, the computer had a glitch. This resulted in frequent failures, and computers were often evaluated by their mean time to failure.

Another type of memory used in early computers was called the "mercury delay line memory." This memory utilized mercury placed in a cylindrical tube and then used so-called pulse trains down the tube. Mercury delay lines had previously been in common use for the precise measurement of time intervals. It took a thirty-two pulse sub-train (package) to represent a binary number with the approximate range of values as a ten-digit decimal number. It took less than ten vacuum tubes to provide the gating and amplification.

A mercury delay line with crystals at the end would store thirty-two 32-bit binary numbers and require very few vacuum tubes. The mercury delay line memory was used in several early computers to improve reliability since this type of memory required fewer vacuum tubes.

The magnetic drum as a storage device was invented around 1900 and was first used in computers by the University of California soon after World War II. The twenty-inch long, eight-inch diameter drum was built using aluminum and was coated with iron oxide, vertically mounted, and rotated at 3,600 revolutions per minute (rpm). Fifty read-write heads were mounted on the cylinder. The drum was divided into ten tracks to the inch, with each track ninety bits to the inch. This resulted in a storage capacity of one-half million binary digits.

Another type of rotating memory was the magnetic disk. The disk was a flat platter approximately twelve to fifteen inches in diameter and was properly coated to become a magnetic storage. There could be a stack of platters with a read-write head for each platter. These read-write heads would move in and out toward the center as required to locate the right track to store or retrieve data.

In the mid-1950s, the magnetic core was developed as a memory device, and it soon became the memory of choice on most computers. Core memory is constructed of doughnut-shaped ferrite material through which three wires are passed in such a manner that zeros and ones can be stored. The wires passing through the little doughnut can magnetize the core in either direction. The core becomes a flip-flop. The designer can then choose the flip for a zero and a flop for the one, or vice versa.

The technology used involves the process of hysteresis loss which Doc Hoelzer and I talked about in 1951. The cores can be arranged in a square array of 32 by 32 cores, which, when multiplied together, gives a product of 1,024—which is generally called 1K in binary where a K (thousand) is 1,024 instead of 1,000.

These square arrays could be stacked to get a memory of 4k thirty-two-bit words. The magnetic core proved to be reliable, relatively cheap, a low user of power and air-conditioning, and soon became the main memory for most computer mainframes built in the late 1950s.

There continued to be improvements in other electronic components such as diodes and transistors. These improvements led to the introduction of printed circuit boards. To make these boards, a circuit diagram was printed on a copper-clad substrate, and then the board was etched—removing the copper not protected by the printing. Electronic components were then placed on the board and dip soldered. Using through-the-board connections, both sides of the board could be used which gave two layers of wiring. These printed circuit boards ended up being about three inches by five inches, giving a good advantage in size, troubleshooting, and cost.

As the magnetic core became the standard random access storage device, there was considerable effort in developing a good system for sequential memory. The magnetic tape had been made faster and more efficient; the magnetic rotating drum had been improved and work was proceeding on the system offering the most hope for the future. This was the rotating magnetic disk with many platters and multiple read-write heads to locate and to store data.

As the decade of the 1950s came to a close, the mainframe computers were considered to be of the second generation. The second-generation computer generally had magnetic-core memory and was constructed using printed circuits with the latest developments in diodes, transistors, and other electronics.

Sequential or backup memory was still predominantly magnetic tape, but other types were beginning to enter the arena of use. For example, UNIVAC had a large reliable storage-drum system named Fastrand. Major computer installations such as large insurance companies established magnetic-tape libraries of hundreds, even thousands, of reels of magnetic tape to store active and archival data. A full-time tape librarian would guard the data and distribute tapes as required.

The first really significant aid to the programmer came with the development of computer programs called "assemblers." These assemblers were the forerunners of what was to be called "programming languages," which were to come along a few years later and were not in common use until the 1960s. Assemblers came into common use in the mid-1950s and continued well into the 1960s.

Generally, the assembler code was translated on a one-to-one basis. A code such as ADD was translated to the binary code according to that required by the computer in use. The program, called the assembler, would make a pass through the program written by the programmer and make the conversion.

Another, and perhaps more important, function of the assembly program was that it made memory location assignments and allowed the program to name these as desired. Up until the advent of the assembly program, the programmer had to keep track of the memory location of each piece of data; now the computer did it.

The assembly program allowed the programmer to place the mnemonic code for the programming steps required on a programming sheet along with the names for data submitted (including the data storage required for each set) and submit these for assembly. This deck, or whatever was used for input, was the program deck. The assembly program would be executed on the program deck and the programmer would be provided back an object deck which was ready to run on the computer. This process was called assembling, or compiling, a program.

Assembly programs were written especially for the computer in use, hence there was no significant improvement in the possibility of transporting a program from one computer to another. In the case of the IBM 704 system, there were two assemblers in common use. One was developed by IBM and was called NYAP11, where the NY stood for New York and the AP for assembly program. Another assembly program in common use on the 704 was called the Share Assembly Program (SAP), and was developed by the group which called itself SHARE, a cooperative group of IBM computer users.

The first programming language to be widely accepted was named FORTRAN (FORmula TRANslation). The first version of the program was completed in 1957 by a group at IBM headed by John Backus. FORTRAN was accepted from the beginning, and the language, which was significantly modified and improved over the years, was in wide use into the 1980s. It took a few years to completely phase out the widespread use of SAP and NYAP11. The transportability of programs written in FORTRAN was still not easy, and the use of FORTRAN compilers was still computer dependent.

FORTRAN was primarily used to solve problems of science and engineering so within a few years (in 1959), a group in the Department of Defense convened to develop a programming language applicable to programs in administrative applications. This group came up with a set of specifications which was closer to the English language than FORTRAN and applied to administrative applications. This language was named COBOL (COmmon Business Oriented Language). It was, like FORTRAN, a great success and in wide use into the 1980s.

In the late 1950s, a group in Europe attempted to develop a programming language which could be used internationally and would be computer independent. It would also combine the characteristics of FORTRAN and COBOL and become a universal computer language. The language was called ALGOL (ALGebraic Oriented Language) and met with some success. There were other endeavors in the 1950s to ease the effort of developing a set of instructions for a computer and to allow computer programs to be transported from computer to computer.

In the late 1950s and early 1960s, the typical second-generation computer had a good operating system (still called the executive or the supervisor) and was programmed using a good assembly language. Some computer installations were moving to FORTRAN, COBOL, or a special language such as a report generator like Report Program Generator (RPG).

Chapter 15

IBM and the Seven Dwarfs

IBM made its entry into the computer business in 1939 by agreeing to build a computer designed by Howard Aiken of Harvard. This computer which has been called the IBM Automatic Sequence Controlled Calculator (ASCC) as well as the Navy Mark I, was designed by Aiken, built by IBM, installed at Harvard, and supported by the navy. Aiken and Harvard took most of the credit for the development of the computer, and as a result there were strained relations between Harvard, the navy, and IBM.

The Mark I was constructed of switches, relays, and rotating shafts, and some said that when the computer was running, it sounded like a room full of ladies knitting socks for their grandchildren. The computer stood eight feet tall, was fifty-one feet long, consisted of 760,000 individual parts, and weighed five tons. The Mark I was not an electronic computer, but is considered by many to be the first working digital computer to be completed in the United States. It was completed in 1944 and did useful work for several years, including some calculations for John von Neumann having to do with the first atomic bomb.

There is little doubt that IBM will be recognized as one of the great corporate stories of the twentieth century. The early success of IBM was due primarily to Thomas Watson Sr. and Thomas Watson Jr., his son. The senior Watson started his career in 1895 by selling cash registers for National Cash Register (NCR) in the bustling city of Buffalo, New York. NCR was run by John Patterson, whose main ambition was to dominate the cash register business and to use whatever tools necessary to accomplish the task. Patterson viewed his competitors as dirty dogs who were to be decapitated, all the while maintaining an outwardly paternal attitude toward his employees. Watson learned from a master, and after a few successful years in Buffalo, he was called to the Dayton, Ohio, headquarters of NCR and given the job of smashing all the competition using whatever means necessary.

In 1912, NCR was hit with an antitrust suit and Watson and Patterson became the first businessmen prosecuted under the newly enacted antitrust laws which were designed to prevent monopolies. NCR was the prime violator and the principal reason for the law being passed. Watson and Patterson were found guilty, fined five thousand dollars, and sentenced to a year in jail. The jail term was later rescinded, but Patterson had already fired Watson in order to shift as much blame as possible.

Watson, unemployed and expecting his first child, landed a job as general manager of a small company called Computing-Tabulating-Recording (C-T-R). Dr. Herman Hollerith, the university professor and researcher who developed the punched-card system to sort, collate, tabulate, and print data for the Census Bureau in the 1880s, had formed a company to build and market his product. The Hollerith Company was later bought by C-T-R. Watson soon gained control of C-T-R and renamed the company the International Business Machines Company (IBM), and as is sometimes said, "The rest is history."

The next computer step taken by IBM, after the unpleasant experience with Harvard, was the IBM 701, which was sometimes called the Defense Calculator. Thomas Watson Jr., then chairman of IBM, had asked the Defense Department what IBM could do to help in the Korean War effort.

The Defense Department responded that a large scientific calculator to solve problems being met in the design of jet aircraft and to help in nuclear physics research would be of the most benefit.

IBM engineers and scientists toured defense and aircraft factories to determine the specifications for the planned computer. As they were going through these facilities and discussing their plans, the IBM engineers received eighteen orders for the computer when it was completed. These orders came when there were no concrete plans for the design of the computer or its capabilities, proving there was a great need for these computers.

When word got back to Thomas Watson Jr., he exclaimed, "We are in the electronic computer business whether we know it or not, and we better move pretty fast!"

Design of the 701 started immediately, and the design was completed in one year. The assembly of the first 701 began in June 1952, and was shipped to the IBM world headquarters at 590 Madison Avenue and installed on March 27, 1953. The IBM 701 was a landmark product and represented the following for IBM:

- the first large-scale electronic computer manufactured in quantity
- the first commercially available scientific computer
- the first computer in which programs were stored in an internal, addressable, electronic memory
- a record-setting time to develop and produce a computer in less than two years from "first pencil on paper" to installation
- a key to transition from punched-card machines to electronic computers
- the first of the pioneering line of the IBM 700 series—the 702, 704, 705, and 709—which IBM built and marketed in the mid-1950s

By the mid- to late 1950s, IBM had a virtual monopoly, and it was estimated they owned between 75 and 85 percent of the market. They were already facing an antitrust suit and other government action to divest them of some of their business. The court case completed in 1956 was a landmark in opening up competition in the computer field.

As the 1950s were ending, IBM continued to dominate the computer business in spite of the fact that some industrial giants had entered the field. Some referred to the contest for computer business as a situation which might be called, "IBM and the Seven Dwarfs." The seven dwarfs were UNIVAC, General Electric, Honeywell, RCA, Burroughs, Bendix, and Control Data Corporation (CDC). All of these manufacturers except CDC intended to take on IBM across the board, to get right in the arena and battle it out. CDC intended to build computers at the high end and build larger and larger computers. The chief designer at CDC was Seymour Cray who was probably the world's best large-scale (supercomputer) designer.

UNIVAC

Remington Rand, which had acquired UNIVAC, did not rest on its premier position in computing but quickly took steps to acquire a budding electronics company which had a computer on the drawing board. The company was Electronic Research Associates (ERA) which was formed by Bill Norris and a group of associates who had served together doing research in the military.

The UNIVAC I used mercury delay lines for memory, and each of the one thousand words of memory was forty-five bits long. The forty-five bits represented eleven decimal digits plus a sign or a memory word which could hold two computer instructions. The computer performed some 465 multiplications per second, a phenomenal speed for 1951. Efforts were made to minimize the use of vacuum tubes by using mercury delay lines for memory, but there were still some five thousand vacuum tubes required and these were kept to a reasonably low failure rate. One customer kept a detailed and accurate account of failures which showed that the central processor was available 81 percent of the time, a good reliability record for computers of that era.

In 1952, the UNIVAC computer received nationwide exposure when it was used to predict the results of the presidential election between Adlai Stevenson and Gen. Dwight D. Eisenhower. Most of the pundits and the news media had predicted a victory for Stevenson, and when the UNIVAC decided early that Eisenhower would be the winner, CBS was afraid to announce the UNIVAC result.

As the evening wore on and the election of Eisenhower seemed more assured, Walter Cronkite announced that the UNIVAC had predicted the Eisenhower win earlier in the evening and the result was held up for fear the UNIVAC had made a mistake.

This incident was the first exposure to computers for most Americans and perhaps gave a much-exaggerated expectation of the power of the computer. With no knowledge of data analysis and statistical prediction methods, the public naturally concluded that the computer had superhuman powers to predict the future, and we were really entering the age of the "giant brains."

A total of forty-six UNIVAC I computers were installed during the period 1951–58 to customers as diverse as the Census Bureau, the Army Map Service, the Prudential Insurance Company, and the Westinghouse Corporation. The computers were doing applications ranging from payroll, sales performance, design of nuclear power reactors, and training musicians.

General Electric

General Electric (GE) made a major commitment to computers, and it appeared from the beginning they would be a major force in the marketplace. GE had worked with the American Banking Association and the Bank of America to get all banks to accept the set of symbols (Magnetic Ink Character Recording, or MICR) developed by the Stanford Research Institute as standard for use on checks and other bank documents.

As a result, GE was given a contract to build the first twenty sets of MICR-handling equipment for the Bank of America and then was allowed to market these systems worldwide. This early history in banking was to put GE in good position with banks when they were ready to market their computers a few years later.

GE selected a site near Phoenix called Deer Valley to be their computer headquarters and manufacturing facility. GE proceeded to build a series of computers with considerable success, and the first, the GE-225, played an important role in the development of computer time-sharing.

A group at Dartmouth University pioneered computer time-sharing using the GE-225. Time-sharing allowed many users to use the computer simultaneously as if it were their own, and for the first time, a user could have complete access to the computer over communication lines. Also at Dartmouth, Professors John Kemeny and Thomas Kurtz developed the BASIC programming language for the GE-225, a tool that simplified programming and allowed students and faculty to do their own programming.

GE built a 400 series computer popular with banks along with the MICR-processing equipment being marketed by GE—making significant inroads into an area previously dominated by IBM. It was a risky move by GE to go head-to-head with IBM, but it was reasoned that if anyone had the resources to compete in the big ring with the big boy, it was GE.

GE also built a large computer numbered the 600 series. The GE 600 series introduced the Symmetric Multi-processing System (SMP) giving the reliability of a redundant system.

The company had sold several 225s to ABMA to be used as laboratory computers. GE challenged IBM across the board in many areas and made major innovations in systems software. Additionally, GE went international in the computer business as they entered cooperative efforts with Bull in Europe and Toshiba in Japan. It appeared GE was in for the long haul and would continue to be a major player.

Honeywell Corporation

The Honeywell Corporation, with its background in electronics and controls, was a natural to go into the computer business. The history of Honeywell in computers is complex due to the number of mergers, acquisitions, and name changes in which it was involved. In 1953, Honeywell acquired full control of Computer Control Corporation (CCC), and in 1955, formed the Datamatic Corporation with the Raytheon Corporation. They announced and built general-purpose computers even though some would have speculated Honeywell would start with control computers.

The Honeywell computers were numbered the 200, 400, and the 800 series which was announced in 1958. Their computers were installed in private and public institutions including the NASA facilities in New Orleans which became part of the Saturn program to land man on the moon.

RCA

RCA, like GE, was prominent in the field of electronics. Unlike some of the other companies, RCA went into the computer business with a splashy announcement and then left as abruptly. As RCA entered the computer business, they hired several brilliant young computer professionals to lead the computer group. RCA proceeded to follow the general pattern and build computers with capability across the board. The computers were numbered 301 and 501, with a larger number on the way. RCA was also involved in the program to land man on the moon when the company was selected to build the automatic checkout computers for the Saturn launch vehicle. This computer, called the RCA110A, however, never met the exacting specifications set by NASA.

BURROUGHS

The Burroughs Corporation was the leader in building bookkeeping machines as well as certain adding machines and calculators. It was natural that they would move toward computers to protect their turf and to expand their business. The most famous machine from Burroughs was called the Sensimatic, a piece of equipment which greatly simplified the basic bookkeeping function of an office.

Burroughs made a major move into the computer business by acquiring a company called Consolidated Electrodynamics located in Pasadena, California. The company was building a computer which became the Burroughs Electrodata 205, to compete with the IBM 650. Also on the drawing boards was a larger computer to be named the Burroughs Electrodata 220. The 205 and the 220 met with moderate success.

Burroughs decided to build a significant computer to be called the Burroughs 5000 (later 5500). The 5000 was built in Paoli, Pennsylvania, home of the Burroughs military products. The 5500 incorporated some unique hardware and software capabilities and introduced a memory design called "memory stacks" which was to provide a more random memory for all parts of the storage. The computer hardware was also designed to make the proposed universal software system (ALGOL) readily installed on the computer. The B5500 met with some success and was installed in several government organizations, including the lunar landing program. Burroughs hung in the computer business longer than most and eventually merged with UNIVAC to form Unisys.

BENDIX

Bendix was another established electronics company which made the move into computers. Bendix announced in the late 1950s it would build a desk-size computer called the G-15. The G-15 was designed by Harry Huskey, who was then at Wayne State University and had been influenced by the design ideas of Alan Turing rather than those of John von Neumann. The G-15 was said to be more difficult to program than the competition, but it was faster and more reliable. The company sold some four hundred of these machines at forty-five thousand dollars each, which led Bendix to announce a larger machine called the Bendix G-20.

The G-20 was planned as competition for the IBM 650. Bendix gave one of these systems to Yale University, where Alan

Perlis was now department head. In return, Perlis was to build an operating system to do for the G-20 what SOAP had done for the IBM 650. It did not turn out as Bendix had hoped, since the time and the technology had changed, and Bendix soon faded as a force in the computer business.

CONTROL DATA CORPORATION

A new company which made major contributions in developing and installing computers starting in the late 1950s was Control Data Corporation (CDC). The company was formed by a break-off group from UNIVAC led by Bill Norris and most of his colleagues from ERA who had joined UNIVAC a few years earlier. Going with Norris to form CDC was Seymour Cray, who was the greatest inventor of large computers of the time.

Part of the deal made with Cray was that he could build a facility some sixty-five miles from Minneapolis at Chippewa Falls on the Chippewa River where he could live and work. For those prospective customers visiting CDC facilities in Minneapolis, it was a real treat to drive out to Chippewa Falls, shake hands with Seymour Cray, and throw a rock into the Chippewa River. To get this treat, a person had to be given a rating as a good prospect to buy a large computer since Cray did not throw his handshakes around foolishly.

CDC built a 3000 series for the low range of users, a 6000 range for the larger computer users, and announced the 7000 series for those who required a computer in the supercomputer range. CDC was very successful for several years and made significant contributions in large-scale computers.

In the 1960s, both GE and RCA exited the computer business. Across the board, pressure from IBM and, in particular, success of the IBM 360, became too great for GE. They sold their entire computer business, including the facilities in Deer Valley, Arizona, to Honeywell. GE had underestimated, as had many other companies, the support required to meet the expectations of the users.

IBM had provided a range of support services which the spoiled users now demanded. Support soon became the tail that wagged the dog and was costing the vendor considerably more than manufacturing costs. There is no doubt but that GE could have continued to build quality and advanced computers, but chose to exit rather than try to meet the ancillary demands.

RCA simply left the business at a regular meeting of its board of directors. This led to a change from the phrase "IBM and the Seven Dwarfs" to "IBM and the BUNCH" (Burroughs, UNIVAC, NCR, CDC, and Honeywell).

Burroughs and UNIVAC merged in the 1980s, forming a company called Unisys, which has all but disappeared as a computer vendor. NCR still builds and services some hardware and software for banks. CDC and Honeywell no longer build computers.

IBM has moved to the area of supercomputers, data management systems, and has sold their personal computer business to a Chinese firm. Cray Corporation still builds supercomputers. Companies like Dell, Apple, and Hewlett-Packard dominate the hardware business and Microsoft, Oracle, and a few others dominate the software business. The advent of personal computers, the Internet, and special computers have once again revolutionized data management and computing.

CHAPTER 16

WHO INVENTED THE COMPUTER?

A question often asked today, nearly seventy years after the whiskey run made by Professor Atanasoff, is, "Who invented the modern computer?" Before this question can be properly answered, a second question must be posed and answered, "What is a computer?" The first question cannot be properly answered without a well-designed and specific answer to the second.

Today, when the word "computer" is used, the definition most commonly assumed is that the computer is electronic, digital (binary), programmable (general purpose), and has the general characteristics of the personal computer which sits on our desk or we carry to business meetings.

Many people will argue and make a good case that John Vincent Atanasoff and Clifford Berry designed (and at least partially built) the first general-purpose electronic digital computer, the ABC.

Each of the words "general purpose," "electronic," and "digital" is important to the definition. The ABC was designed to be electronic, programmable (general purpose), and to use binary arithmetic. The logic planned for the ABC was to be incorporated into computers which came into being a decade later.

The work of Atanasoff and Berry took place during the late 1930s and was interrupted by the onset of World War II. The partially completed ABC was destroyed during the war to make room for another activity. In 1994, a group of students and engineers undertook the task to finish the ABC by rebuilding the computer, and in 1997 they demonstrated a functioning replica.

Simultaneous with the work of Atanasoff and Berry, a brilliant German engineer by the name of Konrad Zuse designed and built a digital computer which he called the Zuse 1. The Zuse 1 was to use binary numbers and the computer and control unit would be separate from the storage (memory). Zuse's computer was not electronic, but used sliding metallic bars to do the logic and arithmetic operations. The bars moved forward and backward to perform the binary operations.

Zuse 1 was not reliable, so Zuse began the process to build an improved computer. He built the Zuse 2 and later completed the Zuse 3 in 1941, which was a reliable computer and performed important computations for the Germans during the war. The Zuse 3 incorporated features such as floating-point arithmetic which placed the Zuse 3 considerably ahead of its time, but it remained nonelectronic.

Howard Aiken of Harvard was getting under way with the design of a computer in the late 1930s. This computer, later known as the Harvard Mark 1, was begun by Aiken in 1937 in connection with his doctoral thesis which required significant computations.

From the beginning, it appeared that the Aiken computer was destined to be electromechanical and not electronic due to Aiken's great disdain for the unreliable vacuum tube (although Aiken was later to claim it was more a matter of dollars than his mistrust of the vacuum tube).

Aiken was able to convince IBM to build the computer if he would furnish the design criteria.

As the war began, Aiken and others were able to convince the navy to fund the operation of the Mark 1 as it was moved to the campus at Harvard. Aiken and some of his staff remained in the navy and operated the computer at Harvard as the Mark 1, providing the only real automatic computer to do large calculations for the U.S. during the war.

John von Neumann and others used the computer to solve some problems involving the nuclear weapons under development, as did others involved in the war effort.

In many discussions, the computer ENIAC (Electronic Numerical Integrator and Computer) is named as the first operating electronic digital computer since it was fully electronic as compared with the electromechanical Zuse and the Harvard Mark 1. The ABC, which met all the criteria of the definition, was not completed.

The ENIAC was built at the Moore School of Engineering at the University of Pennsylvania in Philadelphia. The computer was completed in 1946, too late to help with the war effort but in time to do a lot of important calculations in the late 1940s.

While designing the ENIAC, John Mauchly visited with Atanasoff at Iowa State to discuss computer construction. This visit was to have major consequences a decade or so later when Mauchly was accused of using ideas of Atanasoff in designing the ENIAC, and a major lawsuit resulted. (This legal struggle is fully discussed in the book by Nancy Stern entitled *From ENIAC to UNIVAC*.) The ENIAC did many valuable computations during the late 1940s and is considered to be the predecessor of UNIVAC, the first computer offered for sale in 1951.

The ENIAC was not a stored-program computer: it was necessary to do a rewiring to set up a new program. There is good evidence to support the argument that the Electronic Delay Storage Automatic Computer (EDSAC), built at Cambridge University in England under the direction of Maurice Wilkes, was the first completed and operational stored-program, digital, electronic computer. The EDSAC was finished in 1949 and used mercury delay lines for memory. The memory capacity was thirty-two 18-bit words and the machine executed about six hundred operations a second. The EDSAC introduced a method of stored subroutines to ease the task of programming. Commonly used functions (trigonometric functions, solving simple algebra problems, etc.) would be stored on tape and then could be called by the program when needed. The EDSAC was a successful computer.

The Manchester, England, Small Scale Experimental Computer (sometimes called the Manchester Baby) was completed a year earlier in 1948. It was a fully programmable digital electronic computer. The Manchester Baby was not used as a production computer but as a laboratory computer to test the Williams tube, developed about that time as a type of computer memory.

It is not possible to discuss the first computers without acknowledging the contributions of John von Neumann. One of those rare, genius men, he came from near Budapest in the early twentieth century and became involved with the development of the ENIAC, as well as every other major research effort he could find.

He was a principal in the development of the EDVAC, which improved on the ENIAC in that it would be a stored-program computer. The capability of a stored program is generally added as a requirement in the definition of a computer.

Von Neumann wrote a complete technical description of the EDVAC in 1945 and this paper is considered to be the first complete description of an electronic stored-program digital computer.

There is some disagreement as to whether or not the EDVAC was ever completed. During the period of time that von Neumann was building his first computer, his colleagues at Los Alamos, Nicholas Metropolis and James Richardson, were constructing an identical copy. The computer at Los Alamos was given the name "Mathematical Analyzer, Numerical Integrator and Computer." Von Neumann was quick and happy to point out that this computer would have the mnemonic name MANIAC.

Alan Turing, a genius from England, developed an early interest in computers, and in 1936, wrote a fundamental paper describing what he called the Turing Machine. The paper described how the computer would work and provide a bridge between theoretical mathematics and the process of actually doing something. During the war, Turing worked at the Code and Cypher School at Bletchley Park, which was charged with deciphering the German secret codes.

The researchers at Bletchley Park found a key to a German code and asked the Post Office Department to build a machine to assist with the decoding. The machine, named the Colossus, was engineered by Tommy Flowers of the Post Office with input from Turing and others at Bletchley Park. The Colossus contained all the prerequisites of the electronic digital computer and is considered by some to be the first successful electronic digital binary computer.

Some have called George Stibitz the father of the modern digital computer. In 1939, Stibitz registered with the U.S. Patent Office (patent #2,668,661) what he called the world's first binary computer. It was described as a binary computer with a memory equal to one thirty-second of one kilobyte. The computer consisted of two large racks of electrical relays, eight feet tall, five feet across, and a foot thick.

Stibitz continued to improve and increase the capability of his computer by building models 2, 3, 4, and 5. Model 5 had two kilobytes of memory, some said it resembled a telephone exchange, it filled a very large room, and cost half a million dollars. The computer had floating-point arithmetic and jump (transfer) instructions.

Some have argued that the Stibitz computer was a special purpose device built for a special application (telephone exchange). This would not disqualify it as a digital computer but would raise the question of whether or not it was a general purpose computer.

The analog computer has been around a long time and can be as simple as a cuckoo clock in which wheels turn and at a certain time the bird comes out and sings its song. These early analog devices were all mechanical in nature. Perhaps the best known analog computer in the early 1930s was the Differential Analyzer built by Vannevar Bush in Cambridge, England.

The first electronic analog computer was built by my boss and friend, Dr. Hoelzer. Doc completed the first electronic analog computer while working at Peenemünde on the guidance system for the German V-2, where he made significant contributions to the guidance system of the German missile systems.

He was told on more than one occasion by Dr. von Braun to quit playing with his electronics and go back to his real job. However, as Doc completed his project, Dr. von Braun recognized the value of this development. The electronic analog computer was, of course, much faster in that it measured quantities electronically through voltage levels and resistances. The electronic analog computer was used extensively during the late 1940s and early 1950s before digital computers became commonplace and reliable.

Beginning in the late 1930s, there were efforts in different parts of the world to develop a device to relieve man of the laborious tasks of computing. Many people in Europe, America, and Asia worked independently and essentially were headed down the same path. Perhaps the best answer to the question "Who invented the computer?" would be many people. "Its time had come."

Dr. Hoelzer worked with various companies afterward such as Packard Bell in developing another type of computer called the "hybrid" computer, which combined the good characteristics of the analog computer with the digital computer. During this effort, he became good friends with Max Palevsky, who later formed the successful company Scientific Data Systems, which built a very good time-sharing computer, the SDS Sigma-7.

The digital computer moved ahead so rapidly that the need for and the applicability of the hybrid computer failed to materialize. However, many of the techniques developed were very useful in building flight simulators.

CHAPTER 17

FROM VERSAILLES TO THE MOON SHOT

World War I ended on the morning of November 11, 1918, at precisely eleven o'clock, and the date was designated as Armistice Day. During my childhood, when we reached the eleventh hour of the eleventh day of the eleventh month, everyone stopped work and the company whistle, which could be heard all over town, blew eleven long blasts. All citizens stood, bowed their heads, and said a prayer for all those boys who did not come home but were buried in places like Flanders Field.

That war ended with an armistice, but the details of a peace treaty were determined at a later date. The peace treaty concerning Germany was finalized in June 1919 in Versailles, France, and is thus known as the Treaty of Versailles.

Some of the main terms of the treaty were:

- surrender of all German colonies as the League of Nations mandates
- return of Alsace-Lorraine (sometimes called the breadbasket of Europe) to France
- demilitarization and a fifteen-year occupation of the Rhineland
- a ban on the union of Germany and Austria

- limitation of Germany's army to 100,000 men with no conscription, no tanks, no heavy artillery, no poison gas supplies, no aircraft, and no airships
- limitation of the German navy to vessels under 100,000 tons and no submarines

Germany signed the Treaty of Versailles under protest; the Congress of the United States refused to ratify the treaty. Many people in France and Britain were angry that the trial of the kaiser or the other war leaders did not proceed. The treaty officially ended the war but it did not end the animosity.

Several historians have conjectured that the stringent restrictions of the treaty led to a severe depression for Germany, giving Adolf Hitler an environment which he used to gain power—eventually leading to World War II.

The restriction against airships and aircraft had a major impact on the young German engineers of the early 1920s. They had been excited by stories of the Red Baron and other flight heroes, but they could neither build nor fly airplanes. They chose other ways to join the birds in flight—the glider and the rocket.

Young Germans became world-champion glider pilots, going long distances from launches off slides constructed on steep cliffs of mountains. The formation of groups of young scientists and engineers who experimented with rockets led to the German missile programs. This eventually led to interest in space flight and a trip to the moon fifty years after the signing of the Treaty of Versailles.

During the World War II years, Dr. Wernher von Braun and his team solved many of the basic problems of rocketry and built military rockets, but they never let go of their long-range dreams of space flight.

During the decade of the 1950s, the group worked for the U.S. Army in developing the Redstone Missile (two-hundred-mile range), the solid-propellant Pershing of slightly longer range, and the Jupiter missile of fifteen-hundred-mile range. During the development of the Jupiter, a missile called the Jupiter-C was developed and later used to place the first United States satellite, *Explorer 1*, into orbit. From 1950 to 1956, the group was known as the Guided Missile Development Division.

In 1956, a crash program to develop an Intermediate Range Missile was given to the army, and they formed the Army Ballistic Missile Agency under Maj. Gen. John Bruce Medaris, with the von Braun group as the development team.

In 1958, President Eisenhower formed the National Aeronautics and Space Administration (NASA) with the purpose of exploring space. NASA was formed to absorb the National Advisory Committee on Aeronautics (NACA) with existing centers at Langley, Virginia; Wright Patterson, Ohio; and Moffett Field, California.

The von Braun rocket group was added as another center and was named the George C. Marshall Space Flight Center (MSFC).

The transfer of the von Braun group, comprised of almost forty-seven hundred civilian employees, took place on July 1, 1960. This transfer was a traumatic experience for the army, which felt it had been robbed of a valuable resource which it had nurtured and developed.

In addition to the personnel transferred and the loss of the best rocket team in the world, a corner of the Redstone Arsenal was given to NASA along with a great deal of equipment and many buildings. There were several meetings where tempers flared and voices were raised, but in the end, General Medaris, realizing the move was best for the von Braun group and for the nation, quieted the voices and moved on.

The army was left with the mission to design and build (using contractors) antiaircraft missiles. This led to a major mission in antimissile missiles, which the American media has subsequently named "Star Wars."

The Marshall Space Flight Center became the booster builders for NASA. The Saturn booster was already in the planning stages as the army was developing the capability to place heavier payloads into orbit. The planned mission of the *Saturn* changed when President Kennedy announced plans to send a manned mission to the moon in the decade of the 1960s. That mission is what von Braun had spent his life planning to do, and he had no doubt been influential in the effort to bring President Kennedy to the plan. The president's announcement brought joy and celebration to Dr. von Braun's rocket group.

A jubilant Dr. von Braun announced the plans for the moon journey to his staff at one of the weekly laboratory director staff meetings. He scheduled these staff meetings to start around ten o'clock in the morning, and they sometimes lasted into the early (or even late) evening. The staff meeting announcing the moon shot lasted far into the night.

Dr. von Braun's plan for the moon trip had from the beginning involved a permanent earth-satellite way station where the moon shot would be assembled. The earth-orbiting way station would be permanently manned, and spacecraft to explore the solar system would be assembled on the station as parts were transported from earth. The *Saturn* as originally planned would be capable of serving as the truck from earth to the orbiting space station.

I recalled seeing the von Braun plans for the manned way station in 1951. The manned satellite looked like a large wheel with spokes and a hub at the center where astronauts lived and worked. The wheel would be constantly spinning to maintain stability and to simulate gravity.

Dr. von Braun made a strong pitch for his plan, but in the end the decision was made to go for the moon with a direct shot from earth. It was felt that establishing the assembly earth satellite would add two to three years to the project, which would stretch the project into the next decade. President Kennedy had used the words "the decade of the 1960s" when committing the nation to the moon trip, so the planners felt this was an important part of the goal.

Dr. von Braun thought in the long-term and assumed that after the moon flight, there would be great motivation to go further into space, and the space station would be the jumping-off place for future ventures. He did not think of the moon shot as an end, but rather as a beginning, a first step to explore our solar system.

For the direct moon trip, the *Saturn* needed to be configured and powered to deliver the lunar spacecraft (along with its required propulsion system) into an orbit of the earth so that it could be positioned for the trip to the moon. This required a booster, called a launch vehicle, not yet imagined. It would be necessary to place approximately forty-five tons (ninety thousand pounds as compared with the twenty-eight pounds which we were so proud of just four years earlier) into an earth orbit, and then propel it on its way to the moon.

The payload from earth would include the rocket and fuel to guide the rocket to the moon, orbit the moon, land the spacecraft on the moon with rocket power to leave the moon and rejoin the command module left in moon orbit. In addition, the spacecraft would serve as the workstation and living quarters for the crew. The command module would contain rocket power to come back to earth and controls to position for reentry and landing (splashdown).

To get this forty-five tons into the proper orbit and headed to the moon would require a launch vehicle of several stages, each having multiple engines. Two large stages would be required to move beyond the strong pull of the earth's gravitational forces. The upper (third) stage would include the rocket to place the payload in earth orbit, position it for the flight to the moon, and then send it on its way at twenty-five thousand miles per hour.

Some quick calculations led to the conclusion that the first stage would need to have an earth level thrust of 7.5 million pounds. The largest engine (H-1) available in 1960 gave a thrust of 150,000 pounds. An engine with an order of magnitude greater thrust had to be developed.

Such a rocket motor, with a thrust of 1.5 million pounds, was being planned by the Rocketdyne Division of North American

Left to right: Charles Bradshaw and Dr. Wernher von Braun, circa 1960.

Aviation. This engine, designated as the F-1 engine, stood eighteen feet high and was the largest kerosene-fueled engine under development in the world.

The first stage (S-1C) of the *Saturn V* configured with five F-1 engines would use 4.4 million pounds of liquid oxygen and kerosene during the two-and-a-half minutes of propelled flight. Imagine the pumps, plumbing, and controls to send the more than two thousand tons of fuel into the five engines in those two-and-a-half minutes.

The second stage (S-II) was planned to use five J-2 engines being built and tested by the Rocketdyne Division of North American Aviation. Each of the J-2 engines would develop 200,000 pounds of thrust, giving the second stage a total thrust level of 1 million pounds. The propelled flight of the S-2 would last four minutes.

The third stage would be powered by a single J-2 engine and would need to have stop and restart capability to accomplish the orbit parking and restart to propel the moon craft and the three astronauts to the moon.

The *Saturn* vehicle as finally configured for the moon flight would be longer than a football field (364 feet) and weigh six million pounds. There was a lot of work to be done, and the magnitude of the effort to go to the moon was the biggest engineering project ever undertaken. It involved four hundred thousand engineers and scientists, and thousands more technicians and specialists throughout American industry.

In September 1962, President Kennedy visited Dr. von Braun and the engineers at the Marshall Space Flight Center. Dr. von Braun briefed the president on the status of the *Saturn* development and assured him that the manned mission to the moon could be accomplished in the decade of the 1960s.

At the end of the briefing, Dr. von Braun assembled the nine laboratory chiefs at the static test stand to witness a static firing of the early *Saturn* test (H-1) engines, and to meet President Kennedy. Dr. Hoelzer was unavailable; as his deputy, I was asked to stand in for him. It occurred to me later that Doc may have planned to be absent so that I could meet the president, whom he knew I admired. (Doc remained nonpolitical.)

I had never met a president and was not sure I was up to it. All the other laboratory directors were members of von Braun's original Peenemünde group. I decided to get in the middle of the group and perhaps fit in as a rocket expert.

Dr. von Braun led Kennedy, followed by Vice President Lyndon Johnson and a string of senators and other politicians, down the line of laboratory directors introducing them one at a time.

My first impression of President Kennedy was that he was taller and more handsome than I had thought. He had a good tan and appeared to be in great physical condition. Vice President Johnson was the consummate politician as he sauntered along, greeting and joking with all around him. Senator Robert Kerr of Oklahoma and Governor John Patterson of Alabama followed close behind.

Dr. von Braun introduced his team by saying, "Mr. President, this is Dr. Ernst Stuhlinger who heads our Research Projects Division; this is Dr. Ernst Geissler who heads our Aeroballistics Division; this is Dr. Walter Haeussermann who heads our Guidance and Control Division; this is Mr. Karl Heimburg who heads our Missile Test Division; this is Mr. Charles Bradshaw who heads our Computation Division."

I stood erect and shook the president's hand while saying, "Mr. President, I am pleased to meet you."

The president looked down at me (he appeared to be *much* taller than he was), and in a rather serious tone asked, "What did he say your name was?"

I stood even more erect and replied, "Mr. President, my name is Charles Bradshaw."

The president then asked me another question: "How did you ever get a job down here?"

Many thoughts raced through my mind at the speed of light, and I thought about the day we went to the train station in

Cookeville some ten years earlier, and listened to Homer and Jethro, and how I had seen the Shanks Hotel as we left the train station, and how I had remembered the announcement of Stacey Mott on the radio while listening to Hank Williams sing the song about the wooden Indian.

I wondered where I should start and how much detail I should give the very busy president. As I formulated an answer and looked back up, I saw a big smile on the president's face.

He realized the problem he had given me, so he spoke first. Using his best smile and the Irish twinkle in his eye, he said, "You are doing pretty well for an Englishman."

Dr. von Braun caught the humor of me as the only native English-speaking American in the group, and gave me a big smile and a little pat on the shoulder as the party moved along. All the other important visitors seemed to go through a natural ritual in the bland sort of way they had performed many times before. The president, however, seemed to look everybody in the eye and really get an impression as to what they were all about.

The visit was over and a couple of my German friends kidded me about my name and how it just didn't sound much like that of a rocket scientist. I agreed and hurried on back to my office.

I couldn't wait to get home that evening and call my mom. I wanted to tell her that I had been out in the big world for many years where it was "dog-eat-dog," among the guys that had to "root hog or die," and I had made it. Why, just that day, I stood right in the middle of a group of the most famous rocket scientists in the whole world, chit-chatting with the president of the United States!

The next eight or nine years involved hundreds of companies and some four hundred thousand scientists and engineers all dedicated to placing a man on the moon. The von Braun team proceeded with the formidable task of designing and building the launch vehicle, which was to be named the *Saturn V*. It would deliver the forty-five-ton payload into an earth orbit and place it in a proper attitude for the trip to the moon. Once in the right position, and at precisely the right time, the upper stage of the *Saturn V* would propel the spacecraft and supporting equipment toward the moon at twenty-five thousand miles per hour.

CHAPTER 18

THE SPIRAL FROM VITAL TO EXPENDABLE

A reflection on the fates of Dr. Arthur Rudolph, Dr. Robert Oppenheimer, and Dr. Alan Turing leads one to believe they were utilized by their countries to perform essential tasks and then cast aside when they were no longer needed. Oppenheimer, technical head of the Manhattan Project, was described by Gen. Leslie Groves as perhaps the only man who could have successfully led the building of the atom bomb during the time allotted. Turing has been designated by many computer historians as the father of modern computing, and he was essential to the effort to build a device to break and interpret the German secret code, contributing significantly to winning the Battle of Britain. Both were later maligned and falsely accused of acts bordering on treason.

After the war, Oppenheimer accepted the position as head of the Institute for Advanced Studies at Princeton University, the premier technical and scientific position in the United States. He also served on the Atomic Energy Commission (AEC) which set the country's policy in matters of nuclear devices.

In 1953, President Eisenhower appointed retired Adm. Lewis Strauss to head the AEC. Strauss did not like or trust Oppenheimer, and set about to have him resign. Oppenheimer refused to resign, so in December 1953, Strauss accused Oppenheimer of disloyalty and charged that Oppenheimer had associated with known communists over the previous years—including the time he was technical head of the Manhattan Project.

A board of inquiry was formed to hear the charges. The inquiry provided an opportunity for some ordinary people to embarrass and humiliate an extraordinary man. The FBI had for a considerable period tapped Oppenheimer's phones, but could present no evidence of violations. There was no evidence presented which would lead to conviction.

The fatal blow to Oppenheimer came from a colleague, Edward Teller, who testified that he did not doubt the loyalty of Oppenheimer, but he did not trust his judgment and his leftist leanings. Some have speculated that Teller's testimony was tainted since he was still upset with Oppenheimer for originally opposing the development of the hydrogen bomb proposed by Teller. Some also have since speculated that Teller wanted to be the man with his picture on magazine covers and the subject of headlines which proclaimed him as the father of the hydrogen bomb.

At the end of the hearing, the board of inquiry affirmed the loyalty of Oppenheimer but denied him a security clearance, preventing him from doing further work in nuclear weapons.

Oppenheimer then settled into a quiet life of research as head of the Institute for Advanced Studies at Princeton. Many of his friends felt he was relieved to be free of any responsibility for nuclear weapons development, for which he had always felt some sense of guilt. He bought a sailboat and spent much of his free time sailing in the islands of the Bahamas. Oppenheimer died in Princeton in 1967.

The years after the inquiry were much different for Teller. He did, as many supposed he had planned, become the most prominent nuclear physicist in America and was deeply involved in nuclear weapons development. His name was often in the headlines as he rushed from the classroom at Berkeley to the nuclear facilities at the Livermore Labs and to the Pentagon, and he became known as the father of the hydrogen bomb.

Teller fell from favor with most of his colleagues since they felt Oppenheimer was done an injustice and much of it was Teller's fault. Also, the majority of the nuclear scientists were against developing more powerful nuclear weapons and were making every effort to bring about some international agreement to control nuclear weapons.

In his later years, Teller made attempts to realign with some of his colleagues, but it was too late. Some colleagues noted that the last years of Teller's life were in turmoil while those of Oppenheimer were pleasant and peaceful.

The ideas and theories of Alan Turing were used to build an early, perhaps the first, successful binary-based computer to do logical calculations. This was the Colossus computer used in the British effort to break the German codes. Previously, in 1935–36, Turing had devised what he called the Universal Turing Machine, which was, according to Turing, "an imaginary device which established the possibilities and limitations for every computer which will ever exist, no matter how complex." It would be several years before the Universal Turing Machine would be recognized as the bridge between the old world of logic and the new world where people actually did things.

In 1952, Turing was convicted by a British court for his involvement in "unnatural acts" (homosexuality). He lost his security clearance and was required, instead of going to prison, to take female hormones as a cure for his problem. These hormones led to a change in his body makeup, which caused severe depression, and his life was destroyed. After many months of torture, Turing died by suicide in 1954 by eating an apple he had laced with strychnine.

The moon shot and the development of the atomic bomb are certainly the two most complicated and extensive scientific and engineering events of recorded history to date. Even though both contained science and engineering, the atomic bomb development was more science and the moon shot was more engineering.

There were approximately ten to twelve Nobel laureates intimately involved in the bomb development, and they navigated through unexplored areas of nuclear physics. This is not said to downplay the great engineering feats of building three cities, devices to separate U-235, and two operating nuclear reactors.

The trip to the moon involved designing and building much larger rocket motors, larger and stronger fuel tanks, coordinating the construction by different companies of several rocket stages, and making sure the parts fit together at completion—all of which required expert systems engineering.

The planning for the flight to and landing on the moon, with real live passengers aboard, had to be filled with more unknowns and emotions than at any time since Christopher Columbus set sail for the New World. In working with Dr. von Braun and Dr. Hoelzer, it seemed to me they thought like engineers; while Dr. Householder and Dr. Weinberg at Oak Ridge thought like scientists.

Dr. Helmut Hoelzer retired from NASA in 1973, but he was not finished. He once told me he "re-tired," meaning to put on a new set of wheels and go again. He almost immediately took a job with the Skylab project in Europe and spent three years with that effort.

When he returned to the United States, Doc accepted a consulting position with the Tennessee Valley Authority and later became executive vice president of International Aerospace Technologies in Huntsville.

Doc was given many honors and citations in his later years. At the age of eighty-two, he was asked by the Berlin Museum of Commerce and Technology if they could display his electronic analog computer. Unfortunately, the computer, which he had crated for shipping at Peenemünde and which followed him to White Sands and then to Huntsville, had been cannibalized in the 1970s. The persons doing this did not know the significance of this first electronic analog computer.

The museum wanted to display the computer of Dr. Hoelzer alongside the digital computer (Z3) of Konrad Zuse, which was completed in 1941. Doc was asked to rebuild the original computer and he agreed to the request.

Peter Petroff, a longtime friend and colleague, agreed to finance the project and to supply lab space and equipment at a local company, ADS Environmental Services. Working from memory, a few rough notes, and his doctoral dissertation, Doc set about work on the rebuilding project.

The most difficult task was to find components such as those used in the original computer. Doc was a good scrounger, and he found parts in junkyards, in used computer stores, and in the trunk of his twenty-year-old Mercedes. He completed the project, and an exact duplicate of this first electronic analog computer is now displayed in the Museum of Commerce and Technology in Berlin alongside the digital computer built by Konrad Zuse. Interestingly, Helmut Hoelzer and Konrad Zuse did not meet until the last years of their lives.

The past came back to cause Doc Hoelzer some unpleasantness and disappointment. When he had completed the rebuilding of the first electronic analog computer, he had planned to go to the museum and present the reproduction. The museum director's original request had given him an invitation to speak at the dedication and to give the history of his invention.

As the time neared for the presentation, Doc received a communication from the museum saying that he would not be allowed to speak at the ceremonies—but they would still like for him to send the computer. The indication was given that his alleged association with the Nazi party was the reason for the withdrawal of his invitation to speak.

This was a very disappointing day for Doc and a cruel thing to do to a man in the twilight of his career. He may have been a member of the Nazi party (I never knew and never asked), but I am certain he did nothing during World War II except develop new techniques for controlling a guided missile and to invent the electronic analog computer.

The closest Doc and I came to discussing politics would be occasions when I brought up political issues occurring in the United States about which I was taking a strong stand. Doc refused to discuss them with me or to take a position. Sometimes he would add that he felt all political parties were the same and would be corrupted given enough time in power.

In retirement, Doc and Mrs. Hoelzer spent many happy days touring the back roads and the small towns in southern Middle Tennessee. The towns of Winchester, Huntland, Elora, Sewanee, Cowan, Pelham, Belvidere, Lynchburg, Mulberry, and many others became very familiar to the Hoelzers. They also became acquainted with many of the people in these small towns and the local townspeople looked forward to their next visit from the warm and friendly Hoelzers. The Hoelzers became good friends with Bob and Carol Deutscher, owners of the Crow Mountain Orchard in north Alabama near the Tennessee border. The Crow Mountain Orchard grew the best fruit in the state, or even the world (according to Doc and Mrs. Hoelzer), and the owners loved Dr. and Mrs. Hoelzer.

The Hoelzers found wonderful out-of-the-way places to eat lunch, and on several occasions, I drove down from Nashville to meet them in some small town or at a crossroad. The lunch would consist of home-fried chicken, pork chops, and many vegetables which had been gathered from the garden out back just that morning. There would always be homemade corn bread, buttermilk biscuits, and sliced tomatoes. I heard Doc say, more than once, that lower Middle Tennessee was his favorite place to visit and I wondered if it may have been somewhat like the home of his youth.

Doc died in 1996, and I was asked to give a short eulogy at his funeral. In preparing for this, a lot of things went through my mind. It seemed to me that all the great technical things he had accomplished were overshadowed by what he had meant to so many people who had intersected with him during his lifetime. His technical accomplishments will last throughout history. The life he lived and his friendship meant so much to those who had the good fortune to be called one of Doc's boys or girls. A reunion of these "boys and girls" was held in 1992, when Doc was eighty. Approximately 250 of his former employees and colleagues from throughout the United States spent two days reliving the good old days of the 1950s through the early 1970s. As a eulogy, I tried to express the thanks of all who worked with him for the privilege of his friendship.

During the administration of President Jimmy Carter, an office was established to search out and deport Nazi war criminals who might be living in the United States. The establishment of this office, named the Office of Special Investigations (OSI), in the Justice Department was pushed through Congress by Elizabeth Holtzmann of New York and Joshua Eilberg of Pennsylvania. These two legislators alleged that many war criminals had entered the United States after the war. The von Braun team became one of the early targets of the OSI, as they speculated that among this group of German expatriates, there would be some war criminals to be prosecuted.

The OSI could find only one member of the von Braun group around which there might be a chance to build a case. This one unlucky man was Arthur Rudolph, who had the misfortune of having been given the onerous and hazardous task of building a V-2 production facility, called the Mittelwerk, out of a series of tunnels and excavations under the Harz Mountain.

The German army had supplied the labor force consisting of some German laborers, but more and more prisoners and forced laborers. The working conditions had been horrible; there was barely enough food to keep the workers alive and no doubt many died from lack of nourishment and adequate medical care. Dr. Rudolph himself lost fifty pounds and suffered health problems during this period.

There were worker strikes and sit-downs when the laborers refused to work, and some workers were hanged by the German army. Dr. Rudolph was forced to witness the hangings, as were all the prisoners and German civilians.

Dr. Rudolph was extensively interrogated in 1945 prior to his U.S. service, and in later years, given the highest security clearances and many honors, including the Distinguished Service Award from the president while working with the United States Army and NASA from 1945 to 1969.

Dr. Rudolph retired from NASA in 1969, and in a few years, he and his wife moved to San Jose, California, to be near their daughter who was employed at the NASA Ames Research Center. The Rudolphs were enjoying a pleasant retirement when, in 1982, Dr. Rudolph was accused, but never legally charged, by the OSI of being a war criminal. The alleged crime was mistreatment of the workers and participating in, or perhaps ordering, the hanging of prisoners at the Mittelwerk facility.

The OSI found two of the former prisoners (workers) in Europe who testified they had seen the red-headed Rudolph at the hangings. There was no believable testimony that he had personally mistreated a prisoner, or for that matter, any worker. Dr. Rudolph was described by some of the former workers as a forceful man who required that the prisoners work hard, as he did himself.

Left to right: Chuck Hubbard, Arthur Rudolph, and Charles Bradshaw, 1969.

The Rudolphs were unfamiliar with the law and had never been to a United States court or even had a traffic ticket. They had no good friends in the San Francisco area and did not call their friends in Huntsville for fear of involving them in problems similar to theirs.

No progress was made in resolving the case with the young lawyers from the Office of Special Investigations and, in his frustrated dilemma, Dr. Rudolph was led to accept an offer from the OSI to drop all investigations (criminal charges were never made) if he would give up his U.S. citizenship and leave the country. He would be able to retain the retirement benefits he had earned through some twenty-two years of service to the United States. Upon accepting the deal from the lawyers of the OSI, he was immediately branded a Nazi war criminal.

When their friends in Huntsville learned of the situation, they made many efforts to have Dr. Rudolph and his wife brought back to the United States and given a fair trial on any charges against him. There were some at the national level (including journalist and presidential candidate Pat Buchanan) who strongly believed Dr. Rudolph deserved a fair trial. Buchanan wrote editorials supporting the request from Dr. Rudolph and debated the idea on his television programs.

A major effort to clear the name of Arthur Rudolph was made by William Winterstein, a retired Lt. Colonel who first met the von Braun group in Fort Bliss and later worked with them in the 1960s on the *Saturn* project. Winterstein spent many years and much energy to expose the actions of the OSI, but was never able to work through the bureaucracy and the cover-up. He concluded that Dr. Rudolph had been threatened and harassed over a period of time, eventually signing an agreement to give up his U.S. citizenship and leave the country in order to retain the retirement benefits he had earned over a period of faithful service. The agreement signed by Dr. Rudolph contained confessions that he was responsible for mistreatment of slave laborers, which were not true. Winterstein concluded that the treatment received by Dr. Rudolph was the very antithesis of the American system of justice, which guarantees every citizen a day in court. A book entitled *Gestapo USA—When Justice Was Blindfolded* was authored by Winterstein and published in 2002.

Several scientists, not part of the von Braun team and who did not personally know Dr. Rudolph, decided to make their own investigation. Dr. Friedwardt Winterberg, professor of theoretical physics at the University of Nevada, not only wrote letters but went to Europe and conducted his own investigation. He located persons who had been at the Mittelwerk facility during the time Dr. Rudolph was there, and who testified that Dr. Rudolph had made efforts to get more food and medical help for the prisoners. More than one testified they had never seen Dr. Rudolph mistreat workers. These efforts to bring Dr. Rudolph back for a trial were never successful even though the German government made a full investigation, found no crimes committed by Dr. Rudolph, and restored his German citizenship.

Dr. Rudolph lived out his life in Hamburg, Germany. He died in 1996, a disillusioned man. He was one of the earliest rocket pioneers, designing and building one of the first successful rocket motors in 1931. He made major contributions to the rocket and space flight programs of Germany and the United States. Dr. Rudolph played a major role in developing the Redstone missile, was chief engineer on the Pershing missile, and was head of the *Saturn* Systems Office, leading the development of the *Saturn* launch vehicle which placed the *Apollo* spacecraft in the proper position for its trip to the moon. He did not deserve the treatment he received from his adopted country which he served so faithfully and had learned to love.

On several instances, I had served as an advisor on computers and computing to Dr. Rudolph. In all my associations with him, I saw a hard-driving, success-oriented, kind, and considerate man. The same can be said of the dozens of other members of the von Braun team with whom I had close contact. Men such as Helmut Horn, Werner Dahm, Konrad Dannenberg, Gerhard Heller, Hermann Weidner, Karl Sendler, Erich Goerner, Bernhard Tessman, Otto Hoberg, Hans Gruene, Hans Hosenthien, Oswald Lange, Erich Klauss, Joseph Boehm, Gerhard Reisig, Rudolph Herman, Otto Hirschler, Ernst Steinhoff, and all the laboratory directors and staff scientists, as well as all the team members for whom I worked.

They were all good men, and I have never seen any indication that any of them could be inhumane to their fellow man. They were rocket scientists who drove themselves to succeed and became completely involved in their projects. All of them had great respect for Dr. Wernher von Braun and joined him in his vision of space travel and the exploration of our solar system.

Epilogue

In summary, computers and communication satellites have brought about worldwide changes not imagined in the 1950s. At the same time, space exploration and the utilization of nuclear energy have fallen far short of the hopes and plans of the pioneers.

The von Braun Group

Dr. Hans Friedrich joined Convair (Consolidated-Vultee Aircraft) in San Diego, where he became a colleague of Kraft Ericke and was a principal developer of the Atlas Intercontinental Missile. He died in 1957.

Dr. Friedrich was a kind, competent, and considerate teacher. In later years, when I list the teachers whom I have admired most, he is always near the top. The lessons I learned from Dr. Friedrich went with me through my career and I still occasionally refer to the notes I took under his tutorship. His lectures and assistance put me in good stead for the future years I was to spend working with missiles and launch vehicles, and while teaching courses in differential equations.

In April 1955, Dr. Wernher von Braun and 104 of his associates and their wives became American citizens. The members of the von Braun group then became full-time civil service employees and began building retirement funds and receiving other government benefits. The civil service pay grades did not allow for the proper salary for Dr. von Braun and some of his top scientists when compared to that of comparable private-industry jobs. This seemed of little concern to von Braun and most of the others in comparison with the opportunity they were given to become American citizens, able to proceed with the goal of building rockets and continue with the effort to go into space.

A few left the team due to special circumstances. Due to medical needs, Prof. Theodor Buchold, with whom I carpooled for many months, left to join General Electric for a higher salary. Dr. Adolph Thiel joined TRW as a vice president since he did not believe there was a good future in Huntsville for the visions of Dr. von Braun. Even though the German scientists became American citizens and were free to move where they pleased, only a few left for other jobs. The loyalty to Dr. von Braun and the belief in the mission were paramount to most other considerations.

Professor Hermann Oberth spent a few years in the mid-1950s working with Dr. von Braun in Huntsville, and I was able to visit with him for a few minutes. He was in a large office with Dr. Thiel and two or three others. One of the others in the room was Dr. Erwin Fehlberg, with whom I was later to work in the computation laboratory of Dr. Hoelzer. The contrast between the work technique of Professor Oberth and Dr. Fehlberg was a scene worthy of having crossed the arsenal to witness. Professor Oberth did his formulas in large sweeping and aggressive motions using a roll of butcher paper attached to his desk. At the adjacent desk, Dr. Fehlberg did his formulas very slowly and carefully, almost secretly, on a three-by-five-inch index card.

DR. VON BRAUN

On July 16, 1969, astronauts Neil Armstrong, Mike Collins, and Buzz Aldrin boarded their spacecraft sitting atop the *Saturn V* space vehicle and were rocketed on their way into Earth's orbit and then on to the moon.

Four days later, on July 20, 1969, Armstrong and Aldrin rode the lunar module, *Eagle*, to the surface of the moon at the Sea of Tranquility. Collins remained in the command module, standing by to capture Armstrong and Aldrin for the return to Earth.

The astronauts were in the vicinity of the moon for a day and then headed home for a splashdown in the Pacific Ocean. The three became world heroes, and the event placed the United States firmly ahead in the race into space.

Dr. von Braun had realized his lifelong goal of building a space ship for a trip to the moon; however, there were dark clouds on the horizon. Some in Congress were out to cut the budget of NASA and were asking the question: "What are the benefits to compensate for the billions we have spent to go to the moon?"

Also, it appeared there was an effort under way to break up or even discredit the von Braun team. Dr. von Braun was transferred to NASA headquarters in Washington and placed in a planning office to determine the next NASA venture into space, which most believed would be a manned trip to Mars, our most similar neighboring planet. Some speculated that the move to Washington was not the desire of Dr. von Braun but was forced on him by NASA administrators.

Dr. von Braun soon determined there was no enthusiasm for a national program to take this next step. During his years in Washington, he became very discouraged and looked on with great disappointment as the national leadership began a retreat from space travel and set about to abandon the *Saturn-Apollo* project. Von Braun had spent his life working toward the day when launch vehicles and technical capability would exist to make some bold moves into space. This was a most frustrating time for him.

It would be thirty-five years before someone in authority would recognize the great mistake made in 1969–1971. In October 2005, the newly appointed NASA administrator, Michael Griffen, hinted that he believed NASA had wasted the last thirty years—and many billions of dollars—with the space shuttle and the earth orbiter, when they should have undertaken a bold venture such as a trip to Mars. He encouraged a return to the moon and suggested a permanent manned station on the moon.

Dr. von Braun left NASA in mid-1972 and joined his old friend, Ed Uhl, as vice president of research and development at Fairchild Industries. He maintained residence in Alexandria until his death in 1977.

A simple burial ceremony was held in the Ivy Hill Cemetery in Alexandria. A simple headstone bears his name and the words of Psalm 19:1, which says, "The heavens declare the glory of God; and the firmament showeth his handiwork." This passage was a favorite of Dr. von Braun's from his childhood.

The effort to dismantle the von Braun team became more intense after his death. The Nazi slander leveled against the von Braun group reared its ugly head again, as it had from time to time. Through the years, Dr. von Braun had received many critical letters asking him to explain his activities during World War II. He answered some of them, and stated over and over again, that he had supported his native land during the war effort but he had no knowledge of the Holocaust.

The book entitled *Dr. Space*, written by Bob Ward and published by the Naval Institute Press, contains a letter written by Dr. von Braun to answer the question as to why he did nothing to stop the Holocaust. In this letter, von Braun states that he knew persecutions were going on in Germany, as they were in many countries, but he had no knowledge or suspicion that anything like the Holocaust could take place in Germany. He, like the rest of the world, had no knowledge of the Holocaust until the war was over.

After the Atomic Bomb

Sixty years after the bombs were dropped on Hiroshima and Nagasaki, there are still emotional discussions on whether or not the atomic bombs should have been used. An invasion of the Japanese mainland, without surrender by the emperor, would have certainly resulted in the death of thousands (perhaps hundreds of thousands) of American GIs and surely hundreds of thousands of Japanese soldiers and civilians. As Gen. William T. Sherman said, "War is hell," regardless of whether you win or lose.

I am told that Hiroshima is now a thriving, bustling city and only a few scars remain from the bomb. I would not have believed that possible when I viewed the destruction some two months after the bomb was dropped.

After Germany surrendered in May 1945, the German scientists—including Otto Hahn, Werner Heisenberg, and several others—were rounded up by the British and taken to England. When they heard of the Hiroshima bomb, they did not believe it was a nuclear bomb. They doubted that the Americans had been able to build the bomb which the Germans had not been able to complete. As more evidence came in, the German scientists knew that the Americans had been successful and concluded they could have been first with the bomb if they only had the support of Adolf Hitler and Albert Speer.

Otto Hahn lamented that he felt some responsibility for the deaths of thousands of people and for making the atomic age possible. Later that year he was given the Nobel Prize in physics for his work in leading the effort to split the atom.

The German scientists were allowed to return to Germany with restrictions that they would not go to Russia. There was no need to fear. The nuclear scientists, like the rocket scientists, had no intention of going to Russia as long as Stalin was still in charge.

A review of the American effort to build the atomic bomb in three years' time leads one to believe that this not only represents the greatest scientific accomplishment of all time but also an absolutely gigantic engineering feat. Building the bomb required the creation of three cities in the wilderness, each with all the needed support facilities, and the efforts of some twenty-five thousand people. The engineering required to build the devices for separation of U-235 at Oak Ridge were almost unbelievable. The building of the atomic pile at Hanford to create a chain reaction and produce plutonium was more straightforward, but a big gamble as to the hopes of getting enough plutonium for a series of bombs. The fact that both methods worked is a credit to the scientific genius of the planners and the almost impossible innovation of Gen. Leslie Groves and the builders.

General Groves had a happy and productive life after the war. He retired from the army in 1948; received honorary degrees from many universities including the University of California; and served as a vice president of the Remington Rand Division of the Sperry Rand Corporation. General Groves died in 1970 and received a hero's burial in Arlington National Cemetery.

The Nuclear Age

The growth in the construction of nuclear power facilities was phenomenal during the 1960s and 1970s. In 1979, there were 72 licensed operating reactors in the United States, and 88 more were under construction or in the process of approval; there were over 530 operating worldwide. There are many different types of reactors depending on the fuel used, method of cooling, and the moderators used. The advanced gas-cooled reactor (AGR) uses slightly enriched uranium oxide as fuel, carbon dioxide as coolant, and graphite as the moderator. The pressurized water reactor (PWR) uses enriched uranium oxide as fuel and pressurized water as coolant and moderator. The fast reactor (FR) uses plutonium and uranium oxides as fuel, liquid sodium for coolant, and does not use a moderator. It has more enriched uranium (U-235) and has a smaller core and is constructed of material less absorbent to neutrons. The boiling water reactor (BWR) uses enriched uranium oxide for fuel and water as coolant where the steam from the boiling water passes directly to the turbo generator to create electricity. There are many more designs and configurations of fuel, moderator, and coolant in use or on the drawing boards.

The most exciting promise of energy for the future is from what has come to be known as the breeder reactor. Breeder reactors convert the most abundant form of uranium, U-238, into plutonium-239 (Pu-239), which is similar to the nuclear fuel, U-235. The fast breeder reactor (FBR) is a type of fast neutron reactor that produces more fissile material than it consumes. The Pu-239 breeder reactor is commonly called a fast breeder reactor, and the cooling and heat transfer is done by a liquid metal.

The time it takes for a FBR to produce enough material to fuel a second reactor is called its doubling time. The target time for doubling is ten years. It is possible for an FBR to use its original fuel to produce heat for power generation for ten years and, at the end of that time, have generated enough fuel for another reactor to last ten years. The breeder held out the promise of unlimited resources for energy for the future.

The period from 1955 to 1979 was a period of great optimism for the future growth of nuclear power and its application to many problems. This came to an end due to rumors, mistrust, and the fear of the unknown, primarily by vocal residents of the United States. Japan, France, England, Germany, and other industrialized nations proceeded to solve their energy needs using nuclear energy.

The 1960s and early 1970s saw major changes in the United States in the amount of trust citizens placed in their leaders. One cause that brought about strong emotions was the protest against the use of nuclear energy for any purpose. The vision of the mushroom cloud over Hiroshima and the talk about cancer and deterioration in reproductive organs created a large following for the protest leaders. The building of nuclear power plants became a natural rallying ground for protesters and supporters.

The division in the country over the development of nuclear resources is best exemplified by the following divergent opinions, given during that period, from two persons who should know of what they speak:

"I do not think it is presently worthwhile making a laboratory out of the United States so these people can learn how to deal with reactors."

—Dr. Henry Kendall, professor of High Energy Physics at MIT and chairman of the Union of Concerned Scientists

"Industrial reactors are so safe that we know of not a single individual in America whose health has been damaged by the nuclear nature of such a reactor."

—Dr. Edward Teller, father of the hydrogen bomb and first chairman of the Atomic Energy Commission's committee on nuclear safeguards

The Atomic Energy Commission (AEC) did many studies relative to the dangers associated with nuclear power reactors. One of the most quoted and respected of these reports was AEC number WASH-1400, often referred to as the Rasmussen report. Dr. Norman Rasmussen had calculated, perhaps using some random number theory, that the worst nuclear reactor accident would kill 33,000 people immediately; 45,000 more would die from cancer; and an area of several thousand square miles would be contaminated for years. This result was quoted in many news sources.

The second part of the study was not quoted nearly so often. The second part said the likelihood of an uncontrollable meltdown to cause this result was once in a billion reactor years. A reactor year is one reactor operating for 1 year. A billion reactor years would equal ten thousand reactors operating for 100,000 years.

Today, there are fewer than 500 operating reactors throughout the world. There are only 103 operating in the United States, and these produce 20 percent of the electric power used by the nation. In the 1970s, it was predicted there would be some five times this many operating reactors by the year 2000.

In March 1975, a fire at the Browns Ferry Nuclear Power Station on the Tennessee River near Rogersville, Alabama, became a rallying cry for those opposed to the development of nuclear energy. The accident was caused by a technician searching for air leaks using a candle as the instrument to detect airflow. The resulting fire destroyed several thousand cables, which crippled the emergency core cooling system and resulted in $10 million worth of damage. The Browns Ferry plant was one of the early nuclear facilities of the Tennessee Valley Authority. There was no meltdown or loss of nuclear fuel, and there were no injuries or exposure to radiation by employees or others. Brown's Ferry was taken off line and is just now, in 2007, being readied to come back on line at a cost of some $2 billion.

In 1979, the worst nuclear accident in the United States took place at the Three Mile Island nuclear plant near Harrisburg, Pennsylvania. The reactor was operating at 97 percent capacity when a relatively minor malfunction caused an overheating which led to the automatic opening of a relief valve. It should have closed itself in ten seconds, but it did not. This failure caused a series of events leading to the release of several hundred thousand gallons of cooling water which contained radioactive elements and left 75 percent of the core of the reactor exposed for several hours.

Although there were no injuries due to the Three Mile Island accident, there was great concern over the possibility of radiation-induced health effects, particularly cancer, in the area surrounding the plant. The Pennsylvania Department of Health maintained a registry for eighteen years of more than thirty thousand people who lived within a five-mile radius of the reactor and found no incidences of health problems which could be related to the accident. More than a dozen major independent health studies came to the same conclusion, and in 1996, a federal judge dismissed a class-action suit alleging the accident caused health effects.

There have been similar studies of the after-effects of the bombs at Hiroshima and Nagasaki and the accident at Chernobyl, Russia. The effects of the bombs at Hiroshima and Nagasaki have been much worse. The doses of radiation were much higher, ranging well above the lethal range. Among the survivors, there has been a marked increase in cancer, with the highest rates being among those exposed as children. No evidence of increased genetic effects has been seen in the first and second generation of children born of exposed parents.

The high releases from the Chernobyl accident caused thyroid cancer in children who received high doses from radioactive iodine released in the explosion. However, despite the fact that no health effects have been seen following low-dose releases, this does not dampen the enthusiasm of the protestors nor stem the flow of strict control regulations emanating from state and federal agencies.

The computer facilities at Oak Ridge National Laboratory (ORNL) continued to grow over the years, and the computer group has been at the forefront in advancing technology as relates to scientific research. The installed Cray XT3—which was already one of the fastest computers in the world—has been upgraded to quadruple the speed of the installed computer and will have an operational speed of near one hundred teraflops (a teraflop is one trillion floating-point operations per second), and a computer which measures speed in petaflops (quadrillion flops) is just around the corner. This calculating speed is accomplished by a computer architecture consisting of massively parallel processors (tens of thousands) and memory measured in trillions of bytes.

The presently installed computer has yielded the first three-dimensional numerical simulations of an ignition flame which will help in the design of the next generation of combustion devices. It is hoped the new computer will provide unique and precise views of molecular structure to be used to combat genetic diseases. An appropriation of $55 million has been made available to purchase and install the new computer which will have research users from all parts of the world.

The United States Department of Energy (DOE) recently announced the construction of the best spallation neutron source (SNS) in the world, which has been built at Oak Ridge. "Spallation" is described as what happens when a high-energy proton bombards a heavy atomic nucleus and some of the neutrons are spalled, or knocked out, in a nuclear reaction process.

An SNS is used to bounce a beam of protons off atoms in a crystal and measure the direction and energies at which the neutrons scatter. From a study of these results, it can be determined how the atoms in the crystal are arranged and how they interact. From these results, researchers can be guided in the design and manufacture of better materials for everyday use.

The effort at Oak Ridge is a joint project of all the DOE national laboratories located at Lawrence in Berkeley, California; Los Alamos in New Mexico; Argonne in Illinois; Oak Ridge in Tennessee; Jefferson in Virginia; and Brookhaven in New York. Researchers using this facility will have access to the new computer system being installed at ORNL which will be one of the most capable in the world. These advanced research facilities may, after all, finally make "Dogpatch" the world center of advanced research inside the atom.

Dr. Alston Householder became recognized as a world authority in the area of numerical analysis and accelerating convergence in numerical procedures. His work was described by Prof. Jim Wilkinson in a talk as follows:

> In the 1950s our knowledge of numerical analysis and numerical linear algebra was in a chaotic state. A large number of algorithms had been developed but no systematic study of their relationships had been undertaken. It is primarily due to

> the work of Householder that order has emerged from this chaos. In a remarkable series of papers he effectively classified the algorithms for solving linear equations and computing eigensystems, showing that in many cases essentially the same algorithm had been presented in a large variety of superficially quite different algorithms. The resulting classification made it possible to concentrate on the most profitable lines of research and in this way his work was directly responsible for the development of many of the most effective algorithms in use today. Of particular importance is his appreciation of the value of elementary Hermitian matrices in numerical analysis.

PEOPLE OF THE COMPUTER AGE AND BEYOND

In the period 1969–70, Sam Wyly was convinced that it was much more efficient to transmit data and voice using digital techniques rather than to use analog signals as had been done by the telephone since its invention. In addition, he foresaw the time when most of the information being transferred would be in digital (computer) form rather than voice (analog).

If data were to be transmitted over the analog telephone, it was necessary to convert the digital data from the computer to prepare it for transmission over the analog telephone system. When the analog data arrived at its destination, the reverse conversion had to be made as it entered the digital computer. In our desk computers today, this conversion is accomplished by a small device called the modem (modulate-demodulate).

In 1971, this conversion device was much larger, more expensive, and much slower. At that time, AT&T had a monopoly and there was no competition for long distance transmission, but this did not seem to discourage Wyly.

He decided to form a company to transmit data and voice by digital technique, and he named the company Data Transmission (DATRAN). The company would be partially owned by UCC, where the initial funding would originate. Wyly hired a senior engineer from AT&T and made him president of DATRAN.

During this period of 1970–71, I was employed by UCC and worked for the Government Systems Office based in McLean, Virginia, headed by Jim Hix. Wyly had a longstanding desire to meet Dr. von Braun, so he asked if I could set up a meeting with Dr. von Braun, who was then with NASA in Washington.

In a few days Wyly, Jim, and I met with Dr. von Braun in his NASA office. One of the subjects discussed in the meeting was the plans for DATRAN, and Wyly explained the plans for the new data transmission company. The objective was to start on the East Coast and move toward the West. The larger cities would be connected with high-speed (digital) links over properly positioned microwave towers. In the cities, DATRAN would lease lines already existing and perform the data conversions where required.

Early in the presentation, I saw a slight frown and a question mark come into Dr. von Braun's forehead and eyes. At the opportune time, he asked why the plan was not to use satellites for data transmission. In 1971, communication satellites were not in common use, but Dr. von Braun knew that within a very few years data would be transmitted around the world in just a few milliseconds.

Wyly was in a hurry and probably thought the communication satellites were too far in the future. Today, it is obvious to many that the greatest accomplishment out of the space race of the 1950s is the ability to communicate around the world in fractions of seconds as the data is transmitted in digital form.

That meeting was the last time I saw Dr. von Braun. After a year or so in Washington trying to develop a space exploration plan for the future, Dr. von Braun recognized the obstacles of timing and other priorities of the federal government and finally decided it was time to leave NASA. He accepted a position with his old friend Edward Uhl, head of Fairchild Industries.

When I first learned of this, I remembered one of the jokes Dr. von Braun told in a staff meeting upon returning from a hunting trip with Uhl. Dr. von Braun announced that he had heard of a major merger of companies during his trip, and since we were in a period when a lot of mergers were taking place, we all leaned forward to hear the latest gossip on mergers. Dr. von Braun told us he had learned that Fairchild Industries was planning a merger with the Honeywell Corporation. Some of the details of the merger were still being resolved. However the name of the company had already been selected: *Fairwell Honeychild*!

MORE THAN COMPUTING: ROSS PEROT

The code of conduct and set of principles Ross Perot held for his company obviously carried over to his children. During the years from 1971 to 1988, I was director of the computer center at Vanderbilt University, and two of the Perot children attended the university during this period.

On one occasion, I was asked to fill in for a professor in a computer science class and was told that one of the students in the class was Ross Perot Jr. I asked the professor to describe the son so that I would recognize him and was told I would recognize him without a description.

I assumed he meant he looked a lot like Ross but when I got to class I soon understood. Ross Jr. was the only male in class with a reasonable haircut (better than the one we usually saw on his father) and who was dressed in a clean pair of slacks and a standard-type shirt. He paid careful attention to the lecture and was polite and obviously a young gentleman; I right away concluded that this young man was an eagle.

Ross Perot became politically active during the Nixon administration, particularly as it related to the prisoners of war left in Vietnam. In 1979, two EDS employees were taken hostage by the new Iranian government which had overthrown the shah.

Ross did not sit back and wait for help from the government; he took care of it himself. He immediately organized a rescue mission, headed by Col. Arthur "Bull" Simons, to go to Iran and bring the EDS employees home. EDS employees with experience in the military became part of the mission as did Ross. This rescue mission went to Tehran, and during an uprising in the streets, went into the prison, removed the EDS employees, and brought them home. This mission became the basis for a best-selling novel by Ken Follett entitled *On Wings of Eagles.*

In 1990, Ross became a serious candidate for president, receiving 19 percent of the vote; more than any third-party candidate since Theodore Roosevelt ran on the Bull Moose party ticket.

GEOGRAPHIC CHANGES

The Tennessee Central Railroad continued to provide the highlands area of Middle Tennessee with passenger service until automobiles became more economical and available and the roads became reasonably good. The last trip carrying passengers through Cookeville was made on July 31, 1955. It was a sad day for residents along the way from Harriman to Nashville, who had grown up watching for the trains and setting their daily schedules around that of the trains.

Tennessee Polytechnic Institute was renamed Tennessee Technological University in the 1960s and has grown to an outstanding technical university averaging some ten thousand students. My friend Stacey Mott spent his career working for WHUB and still lives in Cookeville, a dynamic city with optimistic town and civic leaders.

The highlands area, also known as the Cumberland Plateau, is today a popular and ever-growing section of the state. The area is generally bounded by the town of Cookeville on the west, and the town of Wartburg on the east, and moves southwest through the Sequatchie Valley and the towns of Crossville, Monterey, McMinnville, Pikeville, and Dunlap down to Jasper. The area has been described as one of the wildest, prettiest, poorest, and most ecologically diverse regions in the state. Many people choose this area for their homes, since the region is made up of mountains, clear streams, beautiful hardwood forests, and four mild seasons for year-round living.

The city square in Huntsville, Alabama, was completely reworked with a new courthouse and a renovated bank building. The big spring is there, but the rural setting is gone. A beautiful field where watercress once grew is filled with modern buildings including a hotel and a cultural center named for Dr. Wernher von Braun. The city grew from a population of 14,000 in 1951 to 250,000 in 2006. The University of Alabama established a major campus in Huntsville along with an institute for conducting research and offering advanced academic degrees. Many major aerospace firms and other technical companies have facilities in the research park associated with the Research Institute. Nearby is the U.S. Space & Rocket Center, which has on display most of the missiles and space vehicles developed by NASA and the U.S. Army facilities in Huntsville. The missiles on display are full-size and include the *Saturn* launch vehicle which propelled the *Apollo* astronauts to the moon. A good history of our nation's missile and space efforts can be found at the rocket center.

The Marshall Space Flight Center is still a major installation of NASA. Most of the German rocket scientists retired during the 1970s and remained in Huntsville. In 2007, 11 of the original 120 survive and live in Huntsville. They are Werner Dahm, Konrad K. Dannenberg, Hans Fichtner, Dieter Grau, Walter Haeussermann, Heinz Hilton, Oskar Holderer, Walter Jacobi, Rudolf Schlidt, Ernst Stuhlinger, and Georg von Tiesenhausen.

General J. Bruce Medaris retired and became president of the Lionel Company, and we all assumed that Lionel was going to move from electric trains to rockets. Medaris later moved to Florida and became an Episcopal priest. He died in the early 1990s.

The transfer of the von Braun group from the army to NASA in 1960 was a big blow for the army. They still had a major mission in developing antiaircraft/missile technology but no rocket scientists. The army eventually rebuilt its missile capabilities and made major accomplishments in antimissile technology. Programs such as Nike-Zeus and the Patriot missile have been developed and managed from the Army Missile Command located in Huntsville.

Bibliography

Bergaust, Erik. *Reaching for the Stars: A Biography of the Great Pioneer in Space Exploration, Wernher von Braun*. Garden City, NJ: Doubleday, 1960.

———. *Rocket City, USA: Home of the Big Boosters*. New York: Macmillan, 1963.

Bodanis, David. *E=mc²: A Biography of the World's Most Famous Equation*. Winter Springs, FL: ISIS Publishers, 2000.

Ceruzzi, Paul E. *A History of Modern Computing*. Cambridge, MA: MIT Press, 2003.

Clark, John O. E. *The Essential Dictionary of Science*. New York: Barnes and Noble, 2004.

Clark, Ronald W. *Einstein, The Life and Times*. New York: The World Publishing Company, 1971.

Crease, Robert P. *The Prism and the Pendulum: The Ten Most Beautiful Experiments in Science*. New York: Random House, 2003.

DeLamarter, Richard Thomas. *Big Blue: IBM's Use and Abuse of Power*. New York: Dodd, Mead & Company, 1986.

Del Rey, Lester. *Rockets Through Space: The Story of Man's Preparation to Explore Space*. Philadelphia, PA: John C. Winston Co., 1957.

Domenici, Pete V. *A Brighter Tomorrow: Fulfilling the Promise of Nuclear Energy*. Lanham, MD: Rowman & Littlefield Publishers, 2004.

Ford, Kenneth W. *Elementary Particles*. Blaisdell Publishing Company, 1963.

Franklin, Thomas. *An American in Exile: The Story of Arthur Rudolph*. Huntsville, AL: Christopher Kaylor Company, 1987.

Freeman, Marsha. *How We Got to the Moon: The Story of the German Space Pioneers*. Washington, D.C.: 21st Century Science Associates, 1993.

Fritsch, Harald. *Quarks: The Stuff of Matter*. New York: Basic Books, 1983.

Gaines, Matthew. *Atomic Energy*. New York: Grosset and Dunlap, 1970.

Georg, Friedrich. *Hitler's Miracle Weapons: The Secret History of the Rockets and Flying Craft of the Third Reich*. West Midlands, England: Helion and Company, 2005.

Gottfried, Ted. *Alan Turing: The Architect of the Computer Age*. Danbury, CT: Grolier Publishing, 1996.

Gribben, John. *The Scientists: A History of Science Told Through the Lives of Its Greatest Inventors*. New York: Random House, 2002.

Grosch, Herbert. R. J. *Computers: Bit Slices from a Life*. Novato, CA: Third Millennium Books, 1991.

Groves, Leslie R. *Now It Can Be Told: The Story of the Manhattan Project*. New York: Da Capo Press, 1975.

Haeberlin, Scott W. *A Case for Nuclear-Generated Electricity*. Columbus, OH: Battelle Press, 2004.

Householder, Alston. *Principles of Numerical Analysis*. New York: McGraw Hill Book Company, 1953.

Ifrah, Georges. *The Universal History of Computing: From the Abacus to the Quantum Computer.* New York: John Wiley and Sons, 2000.

Johnson, Charles W. and Charles O. Jackson. *Behind the Fence at Oak Ridge*. Knoxville, TN: University of Tennessee Press, 1981.

Johnson, Leland and Daniel Schaffer. *Oak Ridge National Laboratory: The First Fifty Years*. Knoxville, TN: University of Tennesse Press, 1994.

Johnston, Lucile. *The Space Secrets of the Universe*. Birmingham, AL: Roberts and Son, 1969.

Jungk, Robert. *Brighter Than a Thousand Suns*. Orlando, FL: Harcourt Inc., 1958.

Kaku, Michio and Jennifer Trainer. *Nuclear Power: Both Sides*. New York: W. W. Norton and Company, 1982.

Lee, J. A. N. *Computer Pioneers*. Los Alamitos, CA: IEEE Computer Society Press, 1995.

Levenson, Thomas. *Einstein in Berlin*. New York: Bantam Dell-Division of Random House, 2003.

Metropolis, N., J. Howlett, and Gian-Carlo Rota, eds. *A History of Computing in the Twentieth Century*. New York: Academic Press, 1980.

Morris, Robert C. *The Environmental Case for Nuclear Power: Economic, Medical, and Political Considerations*. St. Paul, MN: Paragon House, 2000.

NASA-MSFC Retiree Association. *50 Years of Rockets and Spacecraft in the Rocket City*. Paducah, KY: Turner Publishing Company, 2002.

Neufeld, Michael J. *The Rocket and the Reich: Peenemünde and the Coming of the Ballistic Missile Era*. New York: Simon & Schuster, 1995.

Ordway, Frederick J., III and Mitchell R. Sharpe. *The Rocket Team*. Cambridge, MA: MIT Press, 1982.

Priwer, Shana and Cynthia Phillips. *The Everything Einstein Book*. Avon, MA: Adams Media Corporation, 2003.

Rojas, Raul and Ulf Hashagen, eds. *The First Computers: History and Architecture*. Cambridge, MA: MIT Press, 2000.

Redmond, Kent C. and Thomas M. Smith. *Whirlwind: The History of a Pioneer Computer*. Bedford, MA: Digital Press, 1980.

Rhodes, Richard. *The Making of the Atomic Bomb*. New York: Simon & Schuster, 1986.

———. *Dark Sun: The Making of the Hydrogen Bomb*. New York: Simon & Schuster, 1995.

Rodgers, William. *THINK: A Biography of the Watsons and IBM*. New York: Stein and Day Publishers, 1969.

Russell, Alan K. *The Collector's Book of Science Fiction by H. G. Wells*. Secaucus, NJ: Castle Books, 1978.

Smith, Austin W. *The Story of Tennessee Tech*. Nashville, TN: McQuiddy Printing Company, 1957.

Stanton, Doug. *In Harm's Way: The Sinking of the USS* Indianapolis *and the Extraordinary Story of Its Survivors*. New York: St. Martin Press, 2002.

Stern, Nancy. *From ENIAC to UNIVAC: An Appraisal of the Eckert-Mauchly Computers*. Bedford, MA: Digital Press, 1981.

Ward, Bob. *Dr. Space: The Life of Wernher von Braun*. Annapolis, MD: Naval Institute Press, 2005.

Winterstein, William E. *Gestapo USA*. San Francisco, CA: Robert D. Reed Publishers, 2002.

Index

Page numbers in bold indicate photos and illustrations